Liquid Chromatography
of Polymers and
Related Materials

CHROMATOGRAPHIC SCIENCE

A Series of Monographs

Editor: JACK CAZES

Waters Associates, Inc.
Milford, Massachusetts

Other Volumes in Preparation

Liquid Chromatography of Polymers and Related Materials II

edited by

JACK CAZES
Waters Associates, Inc.
Milford, Massachusetts

and

XAVIER DELAMARE
Waters Associates, Inc.
Paris, France

 **CRC Press**
Taylor & Francis Group
Boca Raton London New York

CRC Press is an imprint of the
Taylor & Francis Group, an **informa** business

First published 1980 by Marcel Dekker, Inc.

Published 2019 by CRC Press
Taylor & Francis Group
6000 Broken Sound Parkway NW, Suite 300
Boca Raton, FL 33487-2742

© 1980 by Taylor & Francis Group, LLC
CRC Press is an imprint of Taylor & Francis Group, an Informa business

First issued in paperback 2019

No claim to original U.S. Government works

ISBN 13: 978-0-367-45203-2 (pbk)
ISBN 13: 978-0-8247-6985-7 (hbk)

Visit the Taylor & Francis Web site at
http://www.taylorandfrancis.com

and the CRC Press Web site at
http://www.crcpress.com

Library of Congress Cataloging in Publication Data

International Liquid Chromatography Symposium,
    Palais des congres et de la musique, 1979.
    Liquid chromatography of polymers and related
materials II.

    (Chromatographic science ; v. 13)
    Includes indexes.
    1. Polymers and polymerization--Analysis--
Congresses. 2. Liquid chromatography--Congresses.
I. Cazes, Jack   II. Delamare, Xavier
            III. Title.
QD139.P6155   1979   547.8'4046   80-16061
ISBN 0-8247-6985-6

PREFACE

Published in this volume are selected papers presented at the
Fourth International Liquid Chromatography Symposium: Liquid Chromatography of Polymers and Related Materials, which was held on Octo-
ber 24-25, 1979 at the Palais des Congres et de la Musique, Strasbourg, France. Included are thirteen papers covering a range of
topics of interest to those involved in the fractionation and characterization of polymeric substances.

We thank the authors of these contributed papers for the fine
work they have done and reported here, and also for their patience
in the preparation of their manuscripts.

Special thanks are extended to Mrs. Cecile Daly for her valuable
assistance, patience, and understanding at all stages of the prep-
eration of this volume.

Lastly, thanks to Waters Associates, Inc., for sponsoring the
symposium and for making their facilities available during the preparation of the final manuscript.

Jack Cazes

Xavier Delamare

CONTENTS

CONTRIBUTORS

K.B. ABBAS, Telefonaktiebolaget LM Ericsson, Stockholm, Sweden

S. ARRIGHETTI, Assoreni, Polymer Research Laboratories, San Donato, Milan, Italy

A.M. BASEDOW, Universitat Heidelberg, Heidelberg, Federal Republic of Germany

R. BRESSAU, BASF Aktiengesellschaft, Kunststofflaboratorium, Ludwig-shafen, am Rhein

B. BRULÉ, Laboratoire Central des Ponts et Chaussees, Paris Cedex, France

M. BRUZZONE, Assoreni, Polymer Research Laboratories, San Donato, Milan, Italy

A. CAMPOS, Universidad de Bilbao, Bilbao, Spain

A.F. CUNNINGHAM, Materials Quality Assurance Directorate, Royal Arsenal East, Woolwich, United Kingdom

J.V. DAWKINS, Loughborough University of Technology, Loughborough, Leicestershire, England

A. DE CHIRICO, Assoreni, Polymer Research Laboratories, San Donato, Milan, Italy

K.H. EBERT, Universitat Heidelberg, Heidelberg, Federal Republic of Germany

J.E. FIGUERUELO, Universidad de Valencia, Burjasot (Valencia), Spain

G.N. FOSTER, Union Carbide Corporation, Bound Brook, New Jersey

Z. GALLOT, Centre de Recherches sur les Macromolecules, CNRS,
    Strasbourg, France

D.J. GOEDHART, Akzo Corporate Research Arnhem, Arnhem, Velperweg 76,
    The Netherlands

A.E. HAMIELEC, McMaster University, Hamilton, Ontario, Canada

C. HEATHCOTE, Materials Quality Assurance Directorate, Royal Arsenal
    East, Woolwich, United Kingdom

D.E. HILLMAN, Materials Quality Assurance Directorate, Royal Arsenal
    East, Woolwich, United Kingdom

J.B. HUSSEM, Akzo Corporate Research Arnhem, Arnhem, Velperweg 76,
    The Netherlands

J. LESEC, Laboratoire de Physico-Chimie des Polymers, Paris, France

T.B. MAC RURY, Union Carbide Corporate, South Charleston, West
    Virginia

J.I. PAUL, Materials Quality Assurance Directorate, Royal Arsenal
    East, Woolwich, United Kingdom

J.P.M. ROELS, The State University of Leiden, Leiden, The Netherlands

B.P.M. SMEETS, Akzo Corporate Research Arnhem, Arnhem, Velperweg,
    The Netherlands

J.A.M. SMIT, The State University of Leiden, Leiden, The Netherlands

V. SORIA, Universidad de Valencia, Burjasot (Valencia), Spain

J.A.P.P. VAN DIJK, The State University of Leiden, Leiden, The
    Netherlands

PROBLEMS ENCOUNTERED IN THE DETERMINATION OF AVERAGE
MOLECULAR WEIGHTS BY GPC VISCOMETRY

J. Lesec

Laboratoire de Physico-Chimie des Polymeres
Paris, France

ABSTRACT

This paper deals with problems that are encountered in modern
GPC with dual detection (refractometer-viscometer) when calculating
average molecular weights, that can lead to incorrect values. Three
main problems are developed: axial dispersion in columns, dependence
of elution volumes upon solute concentration, and application of the
hydrodynamic volume concept. In each case, we have tried to emphasize the best method that takes into account these different effects,
avoiding data treatment errors and that leads, thus, to the best
accuracy in the determination of average molecular weights of polymers.

INTRODUCTION

Gel Permeation Chromatography is a modern method of polymer

characterization which can be used in two different ways. One can

only try to obtain a representation of the polymer molecular distri-

bution. The GPC chromatogram is then directly recorded to compare

polymers with different molecular distributions. Calculations can

then be performed on the chromatographic data to determine average

molecular weight values. In this case, GPC can be considered as a

real measurement system which is going to take the place of tradi-

tional methods of average molecular weight determination and, con-

sequently, as with every technique leading to numerical evaluation

of physical values, must withstand a critical analysis. For this

1

reason, it seems important to take stock of the measurement method's
validity and the accuracy that we can expect from calculations performed with chromatographic data.

We will consider the situation where only the steric exclusion
process arises without perturbing phenomena such as adsorption or
partition whose effects were recently described[1], and we will discuss the different errors that can affect molecular weight values
given by GPC. We can thus classify two kinds of errors[2]:

- typical chromatographic errors
- data treatment errors.

Typical chromatographic errors are directly related to instrumentation and chemicals. They can lead to very wrong molecular
weight values when a chromatographic analysis is not performed under
very well defined conditions. The main reasons are: the molecular
weight accuracy of standards used for calibration, the problem of
sample dissolution, the determination of exact elution volumes in
relation to the mobile phase flow rate stability, the accuracy of
the substance amount which is injected into the column set and peak
shifts resulting from solute concentration and column temperature
effects.

But we will assume here that our chromatographic instrumentation is perfect and we will only examine the problems encountered
in GPC data treatment when calculating polymer average molecular
weights: correction of solute concentration, correction of axial
dispersion and application of universal calibration. We will consider only modern GPC using microgels whose analysis time is about
20 minutes.

PRINCIPLE OF AVERAGE MOLECULAR WEIGHT DETERMINATION

Macromolecular compounds are mainly characterized by a molecular distribution function which depends upon the synthesis process.
Therefore, their physical and mechanical properties are directly related to this distribution which can be discribed by average molecu-
lar weight: $\overline{M}_n$, $\overline{M}_v$, $\overline{M}_w$, $\overline{M}_z$. These molecular weights are usually
determined by classical methods (osmometry, viscometry, light scat-

tering, and ultra-centrifugation). But GPC is the only method which can simultaneously lead to all of these parameters. The principle of chromatogram analysis[3] is represented in Figure 1. It consists in digitizing the refractometric peak into equal fractions, located at the elution volume Vi and whose surfaces are is proportional to solute concentration Ci. The calibration equation Log Mi = f(Vi) gives the relation between Mi and Vi; therefore, summation throughout the peak leads to the average molecular weights.

$$\overline{M}_n = \frac{\Sigma C_i}{\Sigma \frac{C_i}{M_i}} \;,\; \overline{M}_v = \left(\frac{\Sigma C_i M_i^{\alpha}}{\Sigma C_i}\right)^{1/\alpha} \;,\; \overline{M}_w = \frac{\Sigma C_i M_i}{\Sigma C_i} \;,\; \overline{M}_z = \frac{\Sigma C_i M_i^2}{\Sigma C_i M_i}$$

But, if universal calibration[4] Log $[\eta_i].M_i = f(V_i)$ is used, we have to simultaneeously analyse a second chromatogram, given by a viscometric detector[5-7]. In addition, molecular weight distribution curves can be plotted[8], since the relative ratio of each species M_i can be calculated.

In actual fact, the principle of calculation is less evident than it appears. The chromatogram is a very distorted representation of the molecular weight distribution because of the logarithmic nature of the elution [9-10]. If C(M) and W(V), respectively, represent the real distribution and the experimental chromatogram, we can represent the total weight of solute by:

$$\int_{M_1}^{M_2} C(M).dM = \int_{V_1}^{V_2} W(V).dV \tag{1}$$

Considering the general equation of a calibration curve:

$$V = F(M) \tag{2}$$

and its derivative: $F'(M) = \dfrac{dV}{dM}$ (3)

by using (2) and (3), (1) becomes:

$$\int_{V_1}^{V_2} \frac{C(M)}{F'(M)} . dV = \int_{V_1}^{V_2} W(V).dV \tag{4}$$

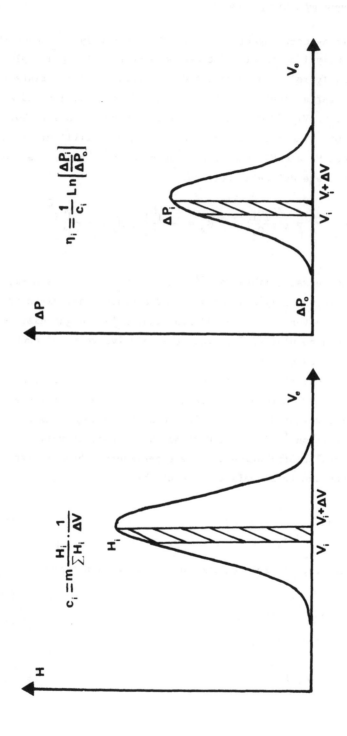

FIGURE 1

Principle of Calculation in Double Detection GPC. (Refractometer-Viscometer). m is the Weight of Solute.

The relationship (4) being true whatever the limits we obtain:

$$C(M) = W(V) \cdot F'(M) \tag{5}$$

In GPC, the logarithmic nature of the elution leads to:

$$W_i(V) = M_i C_i(M) \cdot \frac{1}{Log_e} \cdot \frac{dLogM}{dV} \tag{6}$$

Relationship (6) represents the polymer distribution deformation by the GPC phenomenon and gives a correspondance, point by point, between the chromatogram and the real distribution and, consequently, allows its determination[11].

Fortunately, in the calculation of molecular weights, this deformation is balanced by the detection system which analyses equal volumes. For example, by using (3) and (5), we can write:

$$\int M \cdot C(M) \cdot dM = \int M \cdot F'(M) \cdot W(V) \cdot dM = \int M \cdot W(V) \cdot dV \tag{7}$$

and carrying (1) and (7) in the definition of $\overline{M}_w$, we see that:

$$\overline{M}_w = \frac{\int M \cdot C(M) \cdot dM}{\int C(M) \cdot dM} = \frac{\int M \cdot W(V) \cdot dV}{\int W(V) \cdot dV}$$

The same holds true for $\overline{M}_n$, $\overline{M}_v$, and $\overline{M}_z$. This very interesting result points out that we can calculate an average molecular weight by two different ways, both of which lead to an identical result: either by digitizing in dM the real distribution $C(M)$ or by using directly the chromatogram $W(V)$ digitized in dV, whatever the calibration curve. This demonstration makes the correction of calibration curve slope, included in some authors' calculation methods, not justified.

In fact, the simple method described previously can only be strictly applied with an infinite resolution column set, since molecules M_i, eluted at the elution volume V_i, are contaminated by their neighbors (M_{i-1}, M_{i+1}, M_{i-2}, M_{i+2}, etc...) that have diffused in the mobile phase. This is an axial dispersion phenomenon.

THE AXIAL DISPERSION PHENOMENON

Axial dispersion depends upon the column set efficiency. An abundant literature has been devoted to this effect when column efficiencies are poor. With modern columns, axial dispersion is less important, and we can hope that it will become negligible in the near future. Tung[12] showed that the experimental chromatogram F(V) and the ideal chromatogram W(V) are linked by the general relationship:

$$F(V) = \int_{-\infty}^{+\infty} W(y) \cdot G(v - y) \cdot dy \tag{8}$$

where G(v - y) is the axial dispersion function, which is generally assumed to be guassian[13]. Many solutions were proposed to resolve Equation 8: Fourier transform[14], polynomial methods[12,15], and minimisation method[16]. These methods are not easy to apply and generally require involved computation treatments.

A simple method was recently developed by Marais[17]. Assuming a guassian dispersion function, Equation 8 becomes:

$$F(V_e) = \int_{-\infty}^{+\infty} W(V) \cdot \exp - \frac{(V - V_e)^2}{2 \sigma^2} \cdot dV \tag{9}$$

where V_e is the elution volume of the peak apex and σ the axial dispersion parameter of the guassian dispersion function. A sophisticated resolution of (9) leads to the simple result:

$$\overline{M}_{\beta i} = M_i \left[\exp \frac{\beta \tau^2}{2}\right] \cdot \left[1 + \tau^2 \frac{C'}{C}\right] \tag{10}$$

($\overline{M}_{\beta i} = M_{ni}$ for $\beta = -1$, $\overline{M}_{vi}$ for $\beta = \alpha$, $\overline{M}_{wi}$ for $\beta = 1$) where $\tau = \sigma/a$ (a = slope of the calibration curve), C the distribution curve and C' its derivative. As ratio C'/C is unknown, it can be approximately expressed through H'/H, ratio of the experimental chromatogram derivative on itself:

$$\frac{C'}{C} \approx \frac{H'}{H} \cdot a \tag{11}$$

Relationship 10 permits the calculation of average molecular weights when the parameter τ is determined throughout the chromatogram. The calibration of the parameter σ can be performed by the Waters' recycle method[18] (Fig. 2) which leads to the precise determination of the polydispersity of the polystyrene standards used [19]. We could thus obtain[2] (Fig. 3) a calibration of τ for a

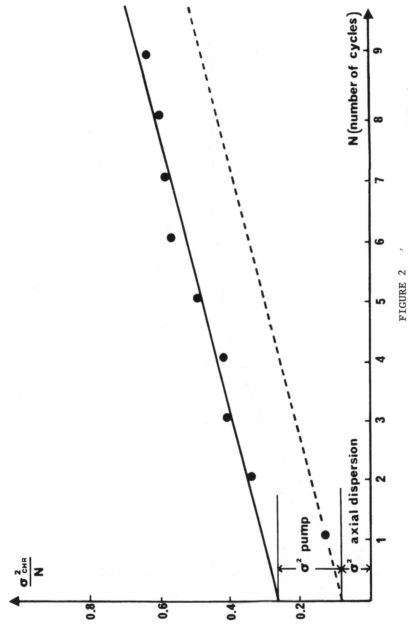

FIGURE 2

Recycling Method of Column Axial Dispersion Calibration and Absolute Polydispersity Determination (Polystyrene Standard 20,500).

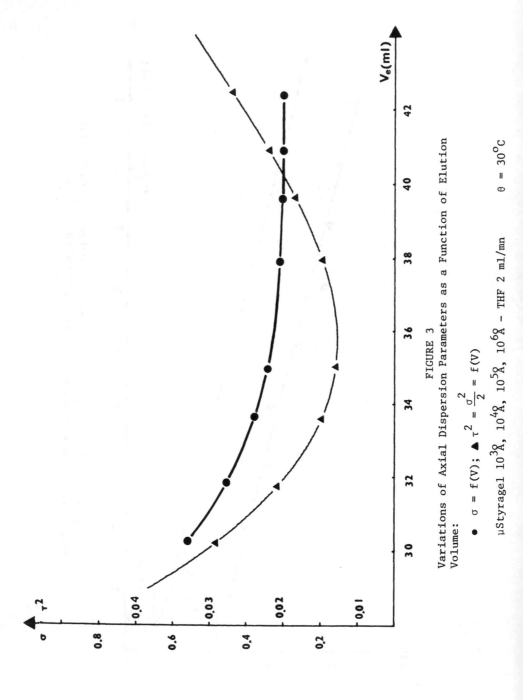

FIGURE 3

Variations of Axial Dispersion Parameters as a Function of Elution Volume:

- $\sigma = f(V)$; ▲ $\tau^2 = \dfrac{\sigma^2}{2} = f(V)$

μStyragel 10^3Å, 10^4Å, 10^5Å, 10^6Å - THF 2 ml/mn $\theta = 30°C$

µStyragel column set. In addition, owing to the high efficiency of
these columns, we could emphasize the first point by contrast with
previous works[18,19]. This point is not located on the same straight
line as the others (Fig. 2) since solute does not pass through the
pump in the first cycle. We could thus point out axial dispersion
in the column set and dispersion in the pumping system[2].

THE SOLUTE CONCENTRATION EFFECT

It should be more exact to speak about segment density since
the experimental observation of this phenomenon is achieved by varying the amount of injected polymer and not the number of moles.
Nevertheless, we will call this phenomenon "concentration effect" as
in the literature. It is well known that increasing sample concentration leads to increasing elution volume[20]. Sometimes, it is
necessary to operate at high concentrations when the detector sensitivity is not sufficient (for example: with a refractometer when
solute refractive index is near that of the solvent or with a viscometer in the low molecular weight range). At high concentrations,
chromatographic peaks are distorted with a skewed front and, unfortunately, these phenomena occur at moderate concentrations with
microgels. Moore[21] explained this distortion by overloading viscosity effects and Rudin[22] developed a theory in which he showed
that hydrodynamic volume depends upon the solute concentration.
This theory is in good agreement with GPC experiments performed in
theta solvents in which such effects are negligible[23]. Finally,
Janca[24,25] pointed out the influence of sample viscosity on chromatographic peak shifting.

To take this effect into account, many solutions have been proposed, Kato[26] suggest the use of a theta solvent for the polymer,
but this solution is not generally easy to apply. Cantow[27] ran
samples at different concentrations, then, extrapolated to zero,
but this method requires several injections. A calibration curve
extrapolated to zero was used by Boni[28], assuming that the concentration effect is negligible for low molecular weight polymers and
that high polymers must be injected at very low concentrations.

The multiple calibration curve method was firstly proposed by Mori[29] who used a set of curves obtained at different concentrations.

In fact, when using microgels, elution volume shifts obviously occur above M = 20,000 (Fig. 4). This effect is very great for high polymers (Fig. 5) and can lead to very significant errors in the calculation of molecular weights when it is not corrected[2]. We propose a correction method which consists of giving an analytical form to the observed effect (Fig. 4)[2]. The calibration function Log M = f(V,C), depending upon the injection concentration, corresponds to an infinite number of calibration curves:

$$\text{Log } M = A_o(C) + A_1(C).V + A_2(C).V^2 + A_3(C).V^3 \tag{12}$$

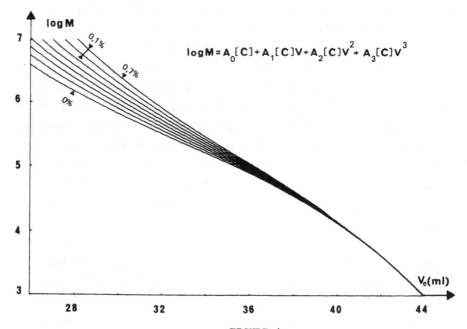

FIGURE 4
Influene of Solute Concentration on Calibration Curve Shape.
μStyragel 10^3Å, 10^4Å, 10^5Å, 10^6Å - THF 2 ml/mn
$\theta = 30°C$

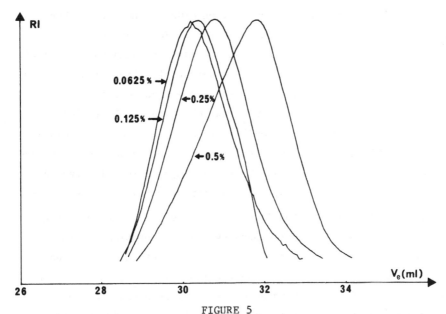

FIGURE 5
Influence of Solute Concentration on Peak Shape and Elution Volume
(Polystyrene Standard 655,000).

μStyragel 10^3Å, 10^4Å, 10^5Å, 10^6Å - THF 2 ml/mn

θ = $30°$C

Each coefficient $A_o(C)$, $A_1(C)$, $A_2(C)$, $A_3(C)$ is itself a polynomial of the third degree which leads to a matrix of 16 coefficients
to define the calibration function. They are give by different polynomial regressions on the data obtained after injections of several
polystyrene standards at different concentrations. This method gives
a very good correction of the concentration effect and permits a precise determination of average molecular weights with an accuracy of
about 5%[7].

THE USE OF HYDRODYNAMIC VOLUME CALIBRATION

It is well known that [η].M is proportional to the macromolecule hydrodynamic volume and is the best parameter for GPC[4].
Generally, the calibration curve Log[η].M = f(V) is used to carry out

analysis of many polymers. However, when GPC molecular weights are
accurately achieved, a difficulty appears. For example, the polydispersity of a 111,000 polystyrene standard is found to be 1.07 with
polystyrene calibration and 1.20 via universal calibration[2]. The
same discrepancy arises when an axial dispersion correction is included in the calculation: 1.03 with polystyrene calibration and
1.07 with universal calibration[2]. This difference can be explained by the effect of axial dispersion on the double detection refractometer-viscometer[7]. Figure 6, where $Log[\eta].M$ is the universal
calibration curve and $Log[\eta]$ the viscosity law from the viscometer,
shows this effect. M value is obtained in dividing $[\eta].M$ by the corresponding $[\eta].M$ by the corresponding $[\eta]$. No problem occurs near
the peak apex, but an error arises when calculation is achieved near
the peak extremities. At a given elution volume, V_e, an incorrect
value of $[\eta].M$ is found because of axial dispersion, whereas viscometry gives a correct value of $[\eta]$. The calculated M is therefore
overestimated on the high molecular weight side and underestimated on
the low molecular weight side; consequently, polydispersity is overestimated. Unfortunately, the same holds true when an axial dispersion correction is applied. The found value of hydrodynamic volume
$[\eta']_i.M'_i$ holds the axial dispersion error $d([\eta_i].M_i)$:

$$[\eta']_i.M'_i = [\eta]_i.M_i + d([\eta]_i.M_i)$$

that is: $[\eta']_i.M'_i = [\eta]_i.M_i + [\eta]_i.dM_i + M_i\,d[\eta]_i$

The calculated M''_i value is therefore:

$$M''_i = \frac{[\eta']_i.M'_i}{[\eta]_i} = M_i + dM_i + M_i \cdot \frac{d[\eta]_i}{[\eta]_i}$$

that is: $M''_i = M_i + dM_i + M_i \cdot d(Ln[\eta]_i)$

Using the Mark-Houwink relationship: $[\eta]_i = k.M_i^{\alpha}$

$$M''_i = M_i + (1 + \alpha)\,dM_i$$

whereas classical calibration gives:

$$M'_i = M_i + dM_i$$

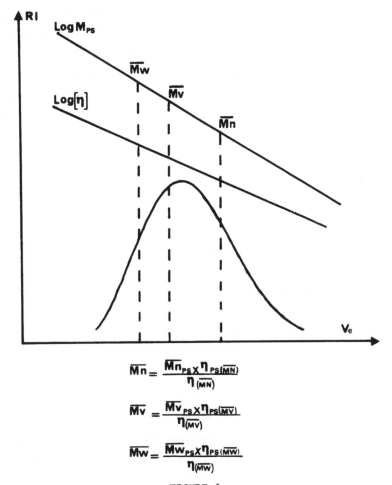

$$\overline{Mn} = \frac{\overline{Mn}_{PS} \times \eta_{PS(\overline{MN})}}{\eta_{(\overline{MN})}}$$

$$\overline{Mv} = \frac{\overline{Mv}_{PS} \times \eta_{PS(\overline{MV})}}{\eta_{(\overline{MV})}}$$

$$\overline{Mw} = \frac{\overline{Mw}_{PS} \times \eta_{PS(\overline{MW})}}{\eta_{(\overline{MW})}}$$

FIGURE 6
Calculation of Molecular Weight through Universal Calibration Curve.

As axial dispersion correction is dM_i, a correct value is obtained through classical calibration, whereas correction is not complete in universal calibration:

$$M''_i - dM_i = M_i + \alpha dM_i$$

Error αdM_i cannot be corrected when α is unknown.

To avoid this drawback and such as hydrodynamic volume concept be used, we suggest the following simple method[2]. First, full calculation, including axial dispersion correction, is performed through a polystyrene calibration function. Correct average molecular weights are thus obtained, but in polystyrene units ($\overline{M}_{n_{ps}}$, $\overline{M}_{v_{ps}}$, $\overline{M}_{w_{ps}}$). Then hydrodynamic volume equalities have merely to be written as shown in Figure 7:

$$\overline{M}_n \cdot [\eta]_{M_n} = \overline{M}_{n_{ps}} \cdot [\eta_{ps}]_{M_n}$$

$$\overline{M}_v \cdot [\eta]_{M_v} = \overline{M}_{v_{ps}} \cdot [\eta_{ps}]_{M_v}$$

$$\overline{M}_w \cdot [\eta]_{M_w} = \overline{M}_{w_{ps}} \cdot [\eta_{ps}]_{M_w}$$

This simple method only requires the viscosity law of polystyrene and the sample viscosity law calculated from the viscometric data: $[\eta]_{M_n}$ and $[\eta_{ps}]_{M_n}$, respectively, are intrinsic viscosities of sample and polystyrene corresponding to the molecular weight $\overline{M}_{n_{ps}}$. The same holds true for $\overline{M}_v$ and $\overline{M}_w$. It is thus possible to obtain average molecular weight values in good agreement with those obtained by other methods[7].

CONCLUSION

As the GPC calculations become more and more accurate, imperfections appear. The first, and the one that is most studied is the axial dispersion effect. Many sophisticated solutions have been proposed but the Marais' simple new method seems to be the easiest to apply and the most reliable. However, axial dispersion is less important as column technology improves and column efficiency increases. In the near future, we can hope that higher efficiency columns will make this correction unnecessary. An important drawback of modern GPC columns is the dependence of elution volumes upon solute concentration. To take this effect into account, we recommend the use of an analytical calibration function which gives a variable calibration curve depending, point by point, upon solute concentration. Finally, the use of universal calibration curve with

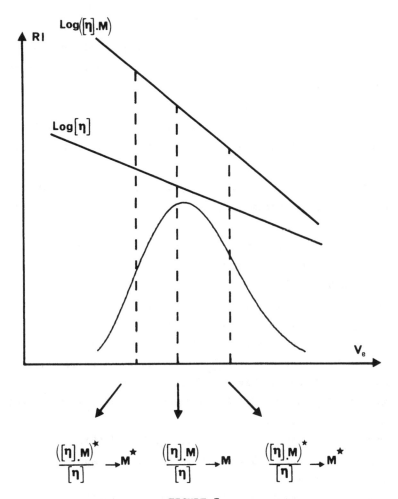

FIGURE 7

Principle of Average Molecular Weight Calculation through Polystyrene Calibration and Sample Viscosity Law.

double detection refractometer-viscometer yeilds a systematic error
leading to an overestimated polydispersity. So that we could,
nevertheless, use the hydrodynamic volume concept, we suggest calculation of average molecular weight values in polystyrene units
through a calssical calibration, then deduction of correct molecular
weight values by writing the equalities of hydrodynamic volumes at
$\overline{M}_n$, $\overline{M}_v$, and $\overline{M}_w$.

REFERENCES

1. Audebert, R., Analusis, 4, 399, 1976; Polymer (in press).

2. Letot, L., Lesec, J., and Quivoron, D., J. Liquid Chromatogr.,
    (to be published).

3. Cazes, J., J. Chem. Educ., 43, 567, (1966).

4. Benoit, H., Grubisic, A., Rempp, P., Dekker, D., and Zilliox,
    J.G., J. Chim. Phys., 63, 1507, (1966); J. Polym. Sci., B5, 753
    (1967).

5. Ouano, A.C., J. Polym. Sci., Symps., 43, 299 (1973). Ouano, A.C.,
    Horne, D.L., and Gregges, A.R., J. Polym. Sci. Phys. Ed., 12, 307,
    (1974).

6. Lesec, J., and Quivoron, C., Analusis, 4, 456 (1976).

7. Letot, L., Lesec, J., and Quivoron, C., J. Liquid Chromatogr.,
    (to be published).

8. Collins, E., Bares, J., and Billmeyer, F., Experiments in Polymer
    Science, John Wiley, New York, 1973.

9. Yau, W.W., and Fleming, W.S., J. Appl. Sci., 12, 2111 (1968).

10. Tung. L.H., Gel Permeation Chromatography, Altgelt, K., and
    Segal, L., Marcel Dekker, New York, 145, 1973.

11. Pickett, H., Cantow, M.J.R., and Johnson, J.F., J. Appl. Polym.
    Sci., 10, 917 (1966).

12. Tung, L.H., J. Appl. Polym. Sci., 10, 375 (1966).

13. Tung, L.H., Moore, J.C., and Knight, G.W., J. Appl. Polym. Sci.,
    10, 1261 (1966).

14. Tung, L.H., J. Appl. Polym. Sci., 13, 775 (1969).

15. Dawidowicz, A., and Sokolowski, S., J. Chromatogr., 125, 428, (1976).

16. Chang, K.S., and Huang, R., J. Appl. Polym. Sci., 16, 329 (1972).

17. Marais, L., Thesis, Strasbourg, 1975. Benoit, H., Marais, L.,
    and Gallot, Z., Analusis, 4, 439 (1976). Marais, L., Gallot, Z.,
    and Benoit, H., J. Appl. Polym. Sci., 21, 1955 (1977).

18. Waters, J.L., J. Polym. Sci., A2, 8, 411 (1970).

19. Grubisic-Gallot, Z., Marais, L., and Benoit, H., J. Polym. Sci.,
    14, 959 (1976).

20. Waters, J.L., Amer. Chem. Soc., Div. Polym. Chem. Prep., 6, 1061,
    (1965).

21. Moore, J.C., Separ. Sci., 5, 723 (1970).

22. Rudin, A., J. Appl. Polym. Sci., 20, 1483 (1976).

23. Kato, Y., and Hashimoto, T., J. Appl. Polym. Sci., 18 1239 (1974).

24. Janca, J., J. Chromatog., 134, 263 (1977).

25. Janca, J., and Pokorny, S., J. Chromatog., 148, 31 (1978).

26. Kato, Y., and Hashimoto, T., J. Polym. Sci. Phys. Ed., 12, 814 (1974).

27. Cantow, M.J.R., Porter, R.S., and Johnson, J.F., Polymer Letters, 4, 707 (1966).

28. Boni, K.A., Sliemers, F.A., and Stickney, P.B., J. Polym. Sci., A2, 1567 (1968).

29. Mori, S., J. Appl. Polym. Sci., 20, 2157 (1976).

HIGH PERFORMANCE GEL PERMEATION
CHROMATOGRAPHY OF POLYMERS

J.V. Dawkins

Department of Chemistry
Loughborough University of Technology
Loughborough
Leicestershire, England

ABSTRACT

The performance of microparticulate GPC packings having narrow particle size distributions has been investigated with short
columns and high speed liquid chromatography instrumentation. Reliable and reproducible data for polydispersity and molecular
weight distribution of polystyrenes have been obtained by optimi-
sing the injection procedure, by using a constant flow pump, and
by incorporating an internal standard into each injected solution.
Column efficiency measurements for a series of polystyrene standards having narrow molecular weight distributions have given
extimates of the chromatogram broadening contributions due to
solute dispersion in the mobile and stationary phases. The
experimental results have been compared with a theoretical expression for polydispersity in terms of eluent flow rate, column packing diameter and solute diffusion coefficient. It is shown that,
for low polymers, mobile phase dispersion is the most important
effect, so that efficient separations may be performed at high
flow rates, whereas for high polymers dispersion owing to mass
transfer into the stationary phase becomes important and it is
no longer possible to achieve efficient separations for high
polymers at fast flow rates. The results for a polydisperse polystyrene demonstrate that the smallest column packings should be
preferred for fast and precise determinations of polydispersity.
The mast accurate determination of molecular weight distribution
is performed at a low flow rate.

INTRODUCTION

It follows from theories for chromatogram broadening that column performance in liquid chromatography is markedly improved by
reducing the particle diameter of the column packing d_p.[1] Microparticulate packings have been produced for separations of large
and small molecules by high performance gel permeation chromatography (HPGPC).[2-4] Particular emphasis has been placed on the
advantages of these small particles packed in short low capacity
columns for high speed (<0.5 hr) separations of polymers and for
high resolution separations of small molecules and oligomers. For
those workers involved in the characterization of high polymers, the
precision of the molecular weight distribution calculated from the
experimental chromatogram is just as important as the speed of separation. However, careful attention to instrumentation and procedures is required in order to obtain accurate molecular weight
data from chromatograms produced by HPGPC with short low capacity
columns.[5,6]

Column efficiency data determined with polystyrene standards
as a function of solute diffusion coefficient D_m, eluent flow
rate u, and d_p will suggest the optimum chromatographic conditions
and are useful in deciding what loss in efficiency and resolution
will occur when separations are performed at high speed. The
importance of obtaining detailed information on column performance
was recognised by Kelley and Billmeyer[7], but their experimental
programme was restricted to crosslinked polystyrene gels with d_p
∿ 50 μm and to inorganic particles with d_p > 100 μm. In this
paper column efficiency results are described for permeating and
non-permeating high polymers with microparticulate packings (d_p ∿
10-20 μm), in order to assess the contributions to chromatogram
broadening in HPGPC arising from solute dispersion in the mobile
phase and during mass transfer. These results show how high
resolution separations may be performed for low polymers and indicate how poor column performance occurs for high polymers at fast

eluent flow rates. It follows that the most precise molecular
weight data for high polymers are obtained at low eluent flow
rates with columns having high efficiencies.

THEORETICAL CONSIDERATIONS

A measure of the efficiency of a chromatography column is the
height equivalent to a theoretical plate or plate height H.[1] The
plate height for an experimental chromatogram is calculated from
the expression

$$H = L/N \tag{1}$$

where L is the column length and N is the plate number which may
be determined from

$$N = 5.54 \left[\frac{V_R}{W_{0.5}} \right]^2 \tag{2}$$

where V_R is retention volume and $W_{0.5}$ is the width of the chromatogram at half its height. Equation (2) is often preferred to the
tangent method[8] because of greater precision. Equation (2) assumes a symmetrical chromatogram corresponding to a normal error
(or Gaussian) function.

In order to interpret the experimental results for plate
height, the dependence of H on the solute dispersion mechanisms
contributing to chromatogram broadening is required. One of the
most widely used relations for predicting column efficiencies, both
in gas chromatography and in liquid chromatography, is the equation
developed by van Deemter and co-workers [9] from a mass balance
approach:

$$H = A + (B/u) + Cu \tag{3}$$

where u is the eluent flow rate, A, B and C are coefficients depending on several parameters (see later), term I (A) is the eddy
diffusion term for solute dispersion in the mobile phase, term II
(B/u) results from dispersion owing to molecular diffusion in the
longitudinal direction in the mobile phase, and term III (Cu) results from solute dispersion owing to resistance to mass transfer.

The plate height may be thought of as the rate of change of peak (or solute zone) variance (in units of length) relative to the distance migrated L.[1] The variance is the square of the standard deviation σ^2, so that H is defined by

$$H = \sigma^2/L \tag{4}$$

If there are several solute dispersion mechanisms contributing to chromatogram broadening, as represented by equation (3), and if these mechanisms are independent of each other, it follows from the laws of statistics that the variance of the chromatogram will be the sum of the variances associated with the individual mechanisms, i.e.

$$H = \Sigma \, \sigma^2/L \tag{5}$$

For a polydisperse polymer, Hendrickson[10] expressed the width of an experimental chromatogram for a permeating polymer in terms of several variables, including the molecular weight distribution of the polymer and chromatogram broadening arising in the column (or columns) from solute dispersion mechanisms. It follows that according to equation (5) the standard deviation σ_o from an experimental chromatogram may be expressed by

$$\sigma_o^2 = \sigma_I^2 + \sigma_{II}^2 + \sigma_{III}^2 + \sigma_M^2 \tag{6}$$

where σ_I, σ_{II} and σ_{III} follow from the solute dispersion terms in equation (3) and σ_M (in units of length) is the standard deviation for the true molecular weight distribution of the polymer. It follows that the experimental plate height is given by

$$H = A + (B/u) + Cu + (\sigma_M^2/L) \tag{7}$$

By analogy with definitions for plate height,[5] we may define σ_M (in units of length) in terms of σ_V (in units of retention volume) with

$$(\sigma_M^2/L) = (L \, \sigma_V^2/V_R^2) \tag{8}$$

where σ_V represents a contribution to the experimental chromatogram. Contributions to equation (6) from chromatogram broadening owing to extra-column effects will be neglected.

A procedure allowing for the polydispersity in the expression for plate height has been described by Knox and McLennan.[11] We shall assume that the true molecular weight distribution of the

polystyrene standards may be represented by a logarithmic normal
distribution, which is reasonable for polymers with narrow molecular weight distributions.[12] For a permeating polymer the
true polydispersity defined as the ratio of the weight average
and number average molecular weights $[\overline{M}_w/\overline{M}_n]_T$ may be calculated
from

$$\ln[\overline{M}_w/\overline{M}_n]_T = \sigma_D^2 \tag{9}$$

where σ_D is the standard deviation in terms of ln molecular weight.[13]
Because the experimental chromatograms for polystyrene standards are
almost symmetrical and because the GPC separation gives an almost
linear calibration plot of log molecular weight versus V_R over the
permeation range, the polydispersity may be calculated from σ_V with
the relation

$$\ln[\overline{M}_w/\overline{M}_n]_T = \sigma_V^2 D_2^2 \tag{10}$$

where D_2 is the slope of the GPC calibration relation between ln
molecular weight M and V_R. With equations (8) and (10) we can show
that equation (7) gives

$$H = A + (B/u) + Cu + (L \ln[\overline{M}_w/\overline{M}_n]_T/D_2^2 \ V_R^2) \tag{11}$$

Since results for permeating and non-permeating solutes as
a function of polymer size will be considered, the coefficients A,
B and C may be written out in full [1], giving

$$H = 2\lambda d_p + (2\lambda D_m/u) + [q \ R(1-R) \ d_p^2 \ u/D_m]$$
$$+ (L \ln[\overline{M}_w/\overline{M}_n]_T/D_2^2 \ V_R^2) \tag{12}$$

where λ (close to unity) is a constant characteristic of the packing,
d_p is the particle diameter of the column packing, λ is a tortuosity
factor, D_m is the solute diffusion coefficient, q is a configuration
factor, which depends on the shape of the pores in the stationary
phase, and R is the retention ratio, defined here for each solute by
V_0/V_R where V_0 is the interstitial (or void) volume of the column
which may be found with a non-permeating polymer. Although equation
(12) may be used as a basis for the interpretation of experimental
chromatograms for polymers having narrow molecular weight distributions, the practical polymer scientist will generally be concerned

with polydisperse samples which will be evaluated in terms of average molecular weights. The experimental polydispersity $[\overline{M}_w/\overline{M}_n]$ may be related to $[\overline{M}_w/\overline{M}_n]_T$ if it is assumed that the chromatogram and the molecular weight distribution are represented approximately by a logarithmic normal function. The experimental value of H is given by $\sigma_o{}^2/L$ in equation (6) and σ_o may be related to $[\overline{M}_w/\overline{M}_n]$ in the same way that σ_M was related to $[\overline{M}_w/\overline{M}_n]_T$ in equations (8), (9) and (10). It follows that equation (12) may be transformed to

$$\ln[\overline{M}_w/\overline{M}_n] + \frac{D_2 V_R{}^2}{L} \{ 2\lambda d_p + (2\gamma D_m/u) + [qR(1-R)d_p{}^2u/D_m] \}$$

$$+ \ln[\overline{M}_w/\overline{M}_n]_T \qquad\qquad\qquad\qquad (13)$$

EXPERIMENTAL

Column Packings

Spherisorb silica S.20.W was supplied by Phase Separations Ltd., Queensferry and three laboratory-prepared macroporous silicas designated H2, H4 and H6 were kindly provided by Dr. J. D. F. Ramsay of AERE Harwell. The silica microspheres (d_p $\sim$15-20 μm) were slurry-packed into individual 316 grade seamless stainless-steel columns (20 cm x 0.3 cm internal diameter). Further details are given elsewhere.[14,15]

Four μStyragel columns each containing a gel of different pore size (in the order of 10^6, 10^5, 10^4 and 10^3 Å, Waters designation) were obtained from Waters Associates, Inc., Hartford, Northwich, Cheshire, England. The gel particles have a diameter of about 10 μm. Each column (30 cm x 0.76 cm internal diameter) was connected in series in the order of decreasing pore size with low dead-volume tubing (5 cm x 0.025 cm internal diameter).

Columns of PLgel 10μ, PLgel 15μ, and PLgel 20μ were kindly provided by Dr. F. P. Warner of Polymer Laboratories Ltd., Church Stretton, Salop, England. These crosslinked polystyrene gel particles have narrow size distributions, as described elsewhere.[16,17] A series arrangement of four GPC columns was used, containing cross-

linked polystyrene gels with exclusion limits of 10^6, 10^5, 10^4 and
10^3 Å. The column dimensions were 30 x 0.77 cm (10μ), 25 x 0.75 cm
(15μ) and 60 x 0.77 cm (20μ).

Materials

Toluene (Analar), tetraphenylethylene (Aldrich Chemical Co.),
and polystyrene (PS) standards (Waters Associates Inc., Pressure
Chemical Co., and Polymer Laboratories Ltd.) were used as received.
The GPC eluent was tetrahydrofuran (BDH Chemicals Ltd.) which was
destabilised, dried, distilled, and degassed before use. The polydisperse polystyrene PSGY2 was prepared by a radical chain polymerisation of styrene at 333 K. The polymerisation was performed
to low monomer conversion (5%) under carefully controlled conditions,
as described elsewhere[6].

Gel Permeation Chromatography

The GPC instrumentation has been described previously.[14,16]
a detailed description of the instrumental requirements and experimental procedures for accurate HPGPC separations of polymers is
given elsewhere.[5,6] HPGPC separations were performed with a
Perkin-Elmer model 1220 positive displacement syringe pump (flow
settings 0.05-6.00 cm^3 min^{-1}, < 3000 lbf/in^2, 500 cm^3 capacity).
In order to ensure a reproducible, constant, pulse-free flow of
the eluent, the pump was initially run at a high flow rate until
the operating pressure was achieved, and the flow rate was then
reduced to that required. With the pumping system at equilibrium,
the retention volume of a solute V_R was calculated from the travel
of the recorder chart paper. In order to increase the reliability
of V_R values, an internal standard, tetraphenylethylene (TPE), was
added to each solution. The retention volume of TPE was assigned
to 100% and the polystyrene standards were measured as a ratio to
this. This method of determining V_R with reference to 100% TPE
was used in the determination of peak heights for the calculation
of average molecular weights.

An Applied Research Laboratories Limited ultra-violet detector
(254 nm, cell volume = 8 μdm^3) was used to detect the solute in the

eluent. A steady baseline, free of noise and drift, was obtained
with the detector on maximum sensitivity. The connecting tubing
between column and detector was shortened in order to minimise
dead volume. Syringe-septum injectors similar to one shown in a
previous paper[16] were used at pressures below 1500 lbf/in^2. For
the silica columns syringe injection (2 μdm^3) through a septum directly onto the top of a column was used.[14] For the crosslinked
polystyrene gel columns, the syringe-septum injection head was
attached directly above the end-fitting of the first column,[16]
so that the solution (injection volume of 40 μdm^3 for polystyrene
standards and small molecules and 80 μdm^3 for polystyrene PSGY2)
could be injected into the solvent flowing into the top of the
column. The injected solutions had solute concentrations in the
range 0.1-0.2% (w/v) with TPE ($\sim$ 0.01%, w/v) added as internal
standard.

    Calculations of molecular weight distribution, number-average
and weight-average molecular weights, $\overline{M}_n$ and $\overline{M}_w$ respectively, and
polydispersity from the chromatograms of the polystyrene standards
and polystyrene PSGY2 were performed by the procedure of Pickett,
Cantow and Johnson.[18] The computer program based on that of
Pickett and co-workers is described elsewhere.[19]

RESULTS AND DISCUSSION

Microparticulate Packings

    Scanning electron micrographs have demonstrated that the silica
particles[14] and crosslinked polystyrene gel particles (PLgel)[16,17]
are regular. Particle size distributions were determined by Coulter
Counter, and number and weight average particle diameters, s_n and s_w
respectively, were calculated with equations presented previously.[16]
It has been proposed that narrow particle size distributions with
$s_w/s_n < 1.2$ should be preferred.[16] Values of this polydispersity
for some of the particles are shown in Table 1, and the spherical
shape and narrow size distribution of the crosslinked polystyrene
gel particles are evident in Figure 1. It was reported previously
that values of H for laboratory prepared silicas were higher than

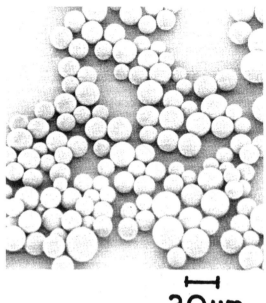

20µm

FIGURE 1
Scanning Electron Micrograph of Crosslinked Polystyrene Gel Particles.

for S.20.W silica,[14] and it has been demonstrated that this behaviour arises from a heterogeneous bed of particles because of
the wider particle size distributions.[15] A non-uniform flow
velocity effect across a column is expected with particles having a
wide size range, because of variable resistance to fluid flow
across a column. Chromatogram broadening owing to solute dispersion
in the mobile phase will then be increased, because miltiple non-
uniform flow paths are generated[7]. A column containing a homogeneous bed of particles is desirable for testing relations for H
such as equation (12). In Figure 2 we present plate height data
for toluene, demonstrating that the optimum performance is expected
for the smallest particles, provided the particles have a narrow
size distribution and can be packed satisfactorily into columns.
From a practical viewpoint, a narrow particle size distribution
minimises the resistance to fluid flow, so that fast separations at
low pressures may be performed. Typical values are shown in Table 1.

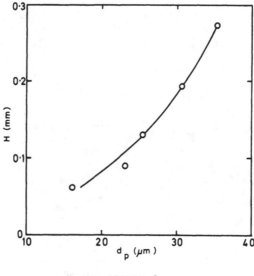

FIGURE 2

Dependence of Plate Height for Toluene of Particle Diameter for
Crosslinked Polystyrene Gel Particles Having Narrow Particle Size
Distribution.

Polymer Dispersion Mechanisms

In order to interpret equation (12) we need to evaluate solute
dispersion in the mobile phase which may be assessed from an experimental study of column efficiency with non-permeating polymers. Such
experiments have been performed for particles having d_p > 40 µm,
suggesting that non-permeating polystyrene standards with M > 10^4
exhibit little or no change in H as u is varied.[7,20-22] In Figure
3, it is observed that the non-permeating polystyrenes PS-110000,
PS-470000, and PS-1987000, which have low values of D_m, give little
variation of H with u. It may be established that these polystyrenes may be regarded as being confined to the mobile phase between
the silica particles (d_p ∿ 20 µm) by inspection of the calibration
curves in reference 14.   Similar behaviour for the dependence of
H on u has been found for PS-1987000 with silicas having larger
pore diameters.[15] We have also shown that non-permeating polystyrenes with molecular weights 111,000, 402,000 and 470,000 have

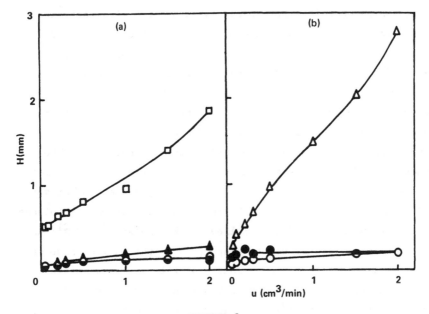

FIGURE 3

Dependence of Plate Height on Flow Rate: (a) S.20.W Silica;● ,
PS-470000 and PS-110000; □ , PS-2350; ▲ , TPE; ○ , Toluene; (b)
H4 Silica; ● , PS-1987000; △ , PS-35000; ○ , Toluene.

TABLE 1

Column Packings and Column Performance at an Eluent Flow Rate of
1 cm^3 min^{-1}.

Packing	s_w/s_n	N (plates m^{-1})	L^a (cm)	Pressure^a drop (lbf/in^2)	Separation^a time (min)
10μ	1.21	30000	120	625	45
15μ	1.15	16000	100	130	35
20μ	1.15	6000	240	160	90
S.20.W	1.11	8200	–	–	–
H4	1.30	5900	–	–	–

^{a} Total length of four columns in series

H values which exhibit little change with u for silica particles
$(d_p \sim 20 \ \mu m)^{24}$, and crosslinked polystyrene particles $(d_p \sim 20$
$\mu m)^{16}$. We may consider the behaviour in Figure 3 in terms of
equation (12). For non-permeating polymers, term III and the
polydispersity term do not arise, and hence H should be a constant
independent of u when term II is negligible. For toluene, there
is no increase in H as u decreases, and term II may be neglected.
The polydispersity terms does not arise for toluene, and so the
dependence of H for toluene on u suggests that the term for solute
mass transfer between the mobile and stationary phases is not very
significant. This is explained by the high value of D_m for toluene
in term III in equation (12). Kelley and Billmeyer [7, 20,21] have
proposed that additional solute dispersion in the mobile phase may
arise from a non-uniform flow velocity profile across a column,
and they interpreted the much higher H values for non-permeating
polystyrene than for a permeating solute as evidence for a velocity
profile effect. In Figure 3, the H values for toluene and the two
polystyrenes are very similar, suggesting that the velocity profile
effect is reduced considerably.

In Figure 3, it is observed that the partially permeating solutes TPE, PS-2350 and PS-35000 exhibit a much more significant
rise in H as u increases than toluene and the non-permeating solutes. A study of permeating solutes covering a wide range of
molecular weight has been performed with the μStyragel columns,
and values of H as a function of u are plotted in Figure 4. It is
clear that apart from the totally permeating solute toluene, and
also polystyrene PS-3600, no minimum in the plot of H versus u
is observed at flow rates above 0.1 $cm^3 \ min^{-1}$. For polystyrenes
with M >10,000, it is apparent in Figure 4 that the slope of each
curve exhibits a trend as the molecular weight of the polystyrene
standard increases. On the other hand the value of H at the lowest or the highest flow rate does not vary in any obvious way with
molecular weight, which may be explained after inspecting equations
(11) and (12) by the observation that the polydispersities of the
six polystyrene standards vary from sample to sample.

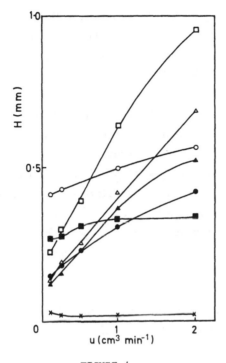

FIGURE 4

Dependence of Plate Height on Flow Rate for µStyragel Columns: □ , PS-470000; △ , PS-200000; ▲ , PS-110000; ● , PS-35000; ○ , PS-9800; ■ , PS-3600; ✕ , Toluene.

A minimum in the plot of H versus u occurs because term II in equation (12) is larger than term III at low u. As the molecular weight of a polymer increases, the value of D_m will decrease so that term II decreases and term III increases. Therefore, term II becomes larger than term III at much lower flow rates as solute size increases, see Figure 4. The absence of minima in the plots in Figure 3 is explained by the low diameter of the silica columns which will have a higher linear flow rate than the µStyragel co-lumns which will have a higher linear flow rate than the µStyragel columns at the same value of u in cm^3 min^{-1} (theory [1] demands that u is interpreted as a linear flow rate).

From the observations in Figures 3 and 4, we may conclude that in practical HPGPC separations for high polymers (M > 10^4)

with low values of D_m term I in equation (12) may be regarded as
a constant independent of u and term II may be neglected. Furthermore, for one polystyrene standard $[\bar{M}_w/\bar{M}_n]$ will be constant. These
observations suggest that the divergence of the curves in Figure
4 as u rises may be interpreted by the dependence of permeation
dispersion on D_m, i.e. term III for mass transfer in equation (12).
Values of $\ln[\bar{M}_w/\bar{M}_n]$ calculated from the chromatograms of the polystyrene standards are plotted versus u in Figure 5. It can be
observed that the relative positions of the curves for the six
polystyrene standards are somewhat different from the positions
in Figure 4. This may be explained by the suggestion that the
polydispersity is calculated from the whole chromatogram whereas
H is deduced from the width of the chromatogram at half-height.
The latter calculation will not allow for chromatogram tails
which are non-Gaussian. The proportion of non-Gaussian tailing
may vary among the polystyrene standards. The dependence of the
slope D_3 of each curve in Figure 5 on molecular weight is apparent

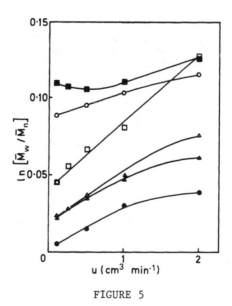

FIGURE 5

Dependence of Polydispersity on Flow Rate for µStyragel Columns:
Symbols as in Figure 4.

for the linear region at flow rates < 1.0 cm^3 min^{-1}. Therefore, the trend observed in Figure 4 is confirmed.

The dependence of permeation dispersion on D_m was demonstrated by plotting in Figure 6 the slope D_3 in Figure 5 determined for the linear part of each curve at flow rates < 1.0 cm^3 min^{-1} for all polystyrene standards except PS-3600. Values of D_m for the polystyrene standards in tetrahydrofuran at 293 K were estimated with the Wilke-Chang equation.[25,26] Error bars are shown for values of Dm as the average error in D_m using the Wilke-Chang equation is 10%.[26] From equation (13), it is clear that V_R^2 and R(1-R) contribute to D_3, but both R(1-R) and V_R decrease as polymer size increases. Therefore, the only explanation for the molecular weight dependence of the slopes of the curves in Figure 5 is the decrease in D_m for larger molecules which will have higher mass transfer dispersion. We may conclude that term III which

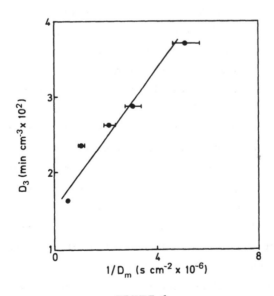

FIGURE 6

Dependence of the Slope of the Plot of Log Polydispersity Versus Flow Rate in Figure 5 on the Reciprocal of the Diffusion Coefficient of Polystyrene.

results from dispersion due to mass transfer is the major contributor to chromatogam broadening for high polymers at fast flow rates.

The results in Figures 4 and 5 have important consequences
for experimental high speed and high resolution GPC separations.
For low polymers, the flow rate may be chosen to give the optimum
column efficiency. At slower flow rates chromatogram broadening
will increase because of term II in equation (3). At faster flow
rates the mass transfer term will not be too significant for low
polymers with high D_m, and so high speed spearations may be performed with little loss in efficiency. Examples of high resolution separations of low polymers are reviewed elsewhere.[4,5] For
high polymers the loss in column efficiency at fast flow rates is
pronounced because of the magnitude of the mass transfer term.
Consequently, the most efficient separations of high polymers
will be performed at low flow rates.

The improvement in column performance as d_p falls in shown
for toluene in Figure 2. The same trend is observed for the polystyrene PSGY2 in Figure 7. Therefore, we may assume that the
polydisperse polystyrene conforms well to equation (13). The GPC
results in Figure 7 demonstrate that the most accurate molecular
weight distribution is obtained with the 10μ gel columns which are
shorter than the 20μ gel columns (see Table 1) and so give faster
separations. The smallest particles are therefore advantageous
for high speed separations. The separation time may be lowered
further by raising the eluent flow rate. For example, in Figure 7,
similar values for $[\overline{M}_w/\overline{M}_n]$ are obtained for the 10μ gel columns
(u = 2 cm^3 min^{-1}) and for the 20μ gel columns (u = 0.5 cm^3 min^{-1})
but the 10μ gel columns give separations eight times faster. The
main advantage of the 20μ gel columns is that they may be used in
low pressure (< 300 1bf/in^2) instrumentation.

The reduction in column performance as u increases results
from the GPC mass transfer mechanism which becomes the dominant
contributor to chromatogram broadening for high polymers at fast
flow rates, see the results in Figure 5. It was also established
in Figure 5 that term II in equation (13) was negligible for poly-

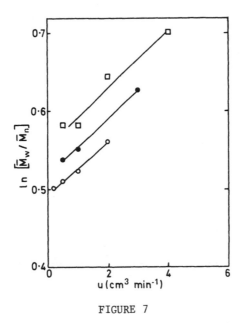

FIGURE 7

Dependence of Polydispersity of Polystyrene PSGY2 on Flow Rate for
PLgel Columns: 0, 10μ gel; ● , 15μ gel; □ , 20μ gel.

styrene standards of high molecular weight ($>10^4$) at practical
flow rates. Therefore, at the lowest flow rate term IV will be a
major contributor to the experimental polydispersity. Polystyrene
PSGY2 was prepared under controlled conditions so that the polydispersity should be close to 1.5[13] because the dominant termination
mechanism in the polymerisation of styrene is radical combination.
[27, 28]   It is evident in Figure 7 that the value of $[\overline{M}_w/\overline{M}_n]$
closest to the polydispersity expected from the polymerisation
mechanism is achieved with the 10μ gel columns at the lowest practical flow rate (u = 0.2 cm^3 min^{-1}) giving a separation time of
225 min.

The differential weight distribution w(M) as a function of
molecular weight for a styrene polymerisation with termination
solely by radical combination is given by[13]

$$w(M) = \frac{4M^2}{\overline{M}_n^{\,3}} \exp\left[\frac{-2M}{\overline{M}_n}\right] \tag{14}$$

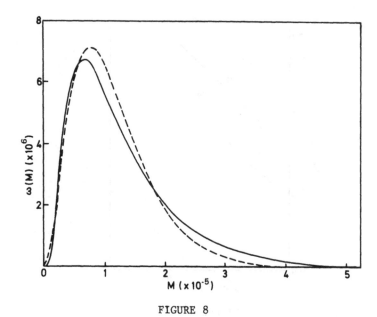

FIGURE 8

Differential Weight Distribution for Polydisperse Polystyrene PSGY2:
———— , from Experimental Chromatogram (μStyragel columns, u =
0.1 cm^3 min^{-1}); ---- , from Equation (14).

This distribution may be calculated for polystyrene PSGY2 whose
value of $\overline{M}_n$ was determined by membrane osmometry and is compared
in Figure 8 with the distribution calculated from the experimental
chromatogram (μStyragel columns, u = 0.1 cm^3 min^{-1}) by the method
of Pickett, Cantow and Johnson.[18] These distributions do not
completely agree over all the molecular weight range, which was
expected because the experimental polydispersity at u = 0.1 cm^3
min^{-1} is 1.61. The difference between this value and the predicted value of 1.5 cannot be explained by neglecting additional termination and transfer reactions in the polymerisation mechanism,
since a distribution function calculated for a termination mechanism in which both radical combination and disproportionation
occur did not provide a better fit of the experimental distribution than the distribution in equation (14).[6] These results show
that chromatogram broadening is still significant at u = 0.1 cm^3

$\min^{-1}$, and term I in equation (13) will still be important even
when term III has been reduced considerably. Therefore studies
of polystyrene PSGY2 with particles having d_p ∿ 5 μm would be
desirable in order to lower chromatogram broadening even further,
when it may be possible to determine precisely an accurate value
of $[\overline{M}_w/\overline{M}_n]_T$ and whether a small fraction of termination by disproportionation does occur.

CONCLUSIONS

These results demonstrate that HPGPC with microparticulate
packings in short columns will provide accurate molecular weight
data when the instrumentation and operating procedures are opti-
mised. A decrease in particle size gives a significant rise in
column performance, and high speed separations at fast flow rates
should be carried out with the smallest particles. The studies of
polydispersity for polystyrene standards show that the extent of
chromatogram broadening will be determined by eluent flow rate and
polymer molecular weight, in agreement with a relation predicted
from the theoretical consideration of solute dispersion mechanisms
in chromatographic columns. The similarity of plate height values
for non-permeating polystyrenes and toluene suggests that mobile
phase dispersion is the major cause of chromatogram broadening for
small molecules. Because mass transfer dispersion is quite low for
small molecules, high resolution GPC separations of oligomers and
low polymers may be performed at fast flow rates. High performance
will not, however, be obtained from microparticulate packings if
the particles are not regular and do not have a narrow size distribution. Mass transfer dispersion becomes much more important for
high polymers and increases as polymer diffusion coefficient decreases and therefore as molecular size increases. Consequently,
extensive chromatogram boradening will occur for permeating high
polymers at fast flow rates, the decrease in column efficiency
with flow rate becoming more significant as molecular size rises.

Efficient separations of high polymers are only obtained at extremely low flow rates when mass transfer dispersion is much reduced. The most accurate determination of polydispersity and molecular weight distribution for a polydisperse polymer should be performed at low eluent flow rates with columns containing the smallest particles, e.g. 10μ gel. From a comparison of experimental and theoretical distributions for a polydisperse polystyrene sample, a small chromatogram broadening contribution would appear to be present even at the lowest practical flow rate of $0.1 \text{ cm}^3 \text{ min}^{-1}$.

ACKNOWLEDGEMENTS

This paper is a review of some of the research carried out on high performance gel permeation chromatography at Loughborough University of Technology during the past five years. The author wishes to thank his colleagues Graham Taylor, Tadeusz Stone, Graham Yeadon and Frank Warner for their contributions to the research described in this paper. The help of Mr. R. E. Buxton, Mr. D. Pinder and Mr. M. Hayles in the characterization of the column packings is gratefully acknowledged. The author is indebted to the following for helpful discussions: Mr. L. J. Maisey, Polymer Supply and Characterisation Centre, RAPRA; Mr. J. S. Hobbs, Applied Research Laboratories; Dr. J. D. F. Ramsay, Dr. R. L. Nelson, Dr. D. C. Sammon and Mr. M. J. Holdoway, AERE Harwell. The research in this paper was supported by grants from Applied Research Laboratories and the Science Research Council.

REFERENCES

1. Giddings, J. C., Dynamics of Chromatography. Part 1: Principles and Theory, Marcel Dekker, New York, (1965).

2. Majors, R. E., J. Chrom. Sci., 15, 334 (1977).

3. Vivilecchia, R. V., Lightbody, B. G., Thimot, N. Z. and Quinn, H. M., J. Chrom. Sci., 15, 424 (1977).

4. Krishen, A., J. Chrom. Sci., 15, 434 (1977).

5. Dawkins, J. V. and Yeadon, G., in Developments in Polymer Char-
    acterisation - 1, (Ed. J. V. Dawkins), Applied Science Publish-
    ers, London, (1978), Chap. 3.

6. Dawkins, J. V. and Yeadon, G., Polymer, in press.

7. Kelley, R. N. and Billmeyer, F. W., Separation Sci., 5, 291 (1970).

8. Grushka, E., Snyder, L. R. and Knox, J. H., J. Chrom. Sci., 13,
    25 (1975).

9. van Deemter, J. J., Zuiderweg, F. J. and Klinkenberg, A., Chem.
    Eng. Sci., 5, 271 (1956).

10. Hendrickson, J. G. J. Polym. Sci. Part A-2, 6, 1903 (1968).

11. Knox, J. H. and McLennan, F. Chromatographia, 10, 75 (1977).

12. Dawkins, J. V., Maddock, J. W. and Coupe, D., J. Polym. Sci.
    Part A-2, 8, 1803 (1970).

13. Peebles, L. H., Molecular Weight Distributions in Polymers,
    Wiley-Interscience, New York, (1971).

14. Dawkins, J. V. and Yeadon, G., Polym. Preprints, 18(2), 227
    (1977).

15. Dawkins, J. V. and Yeadon, G., J. Chrom., submitted.

16. Dawkins, J. V., Stone, T. and Yeadon, G., Polymer, 18, 1179
    (1977).

17. Dawkins, J. V., Stone, T., Yeadon, G. and Warner, F. P., Poly-
    mer, in press.

18. Pickett, H. E., Cantow, M. J. R. and Johnson, J. F., J. Appl.
    Polym. Sci., 10, 917 (1966).

19. Croucher, T. G., Ph.D. Thesis, Loughborough University of Tech-
    nology, (1976).

20. Billmeyer, F. W. and Kelley, R. N., J. Chrom., 34, 322 (1968).

21. Kelley, R. N. and Billmeyer, F. W., Anal. Chem., 41, 874 (1969).

22. Giddings, J. C., Bowman, L. M. and Meyers, M. N., Macromole-
    cules, 10, 443 (1977).

23. Dawkins, J. V. and Taylor, G., Polymer, 15, 687 (1974).

24. Dawkins, J. V. and Taylor, G., J. Polym. Sci. (Polym. Letters
    Edn.), 13, 29 (1975).

25. Wilke, C. R. and Chang, P., Am. Inst. Chem. Eng. J., 1, 264
    (1955).

26. Reid, R. C. and Sherwood, T. K., The Properties of Gases and
    Liquids, McGraw-Hill, New York, (1965), Chap. 8.

27. Berger, K. C. and Meyerhoff, G., Makromol. Chem., 176, 1983
    (1975).

28. Berger, K. C., Makromol. Chem., 176, 3575 (1975).

INVESTIGATIONS CONCERNING THE MECHANISM OF GPC

A. M. Basedow and K. H. Ebert

Institut fur Angewandte Physikalische Chemie
Universitat Heidelberg
Heidelberg, Federal Republic of Germany

ABSTRACT

The accurate determination of molecular weight distributions (MWD) is nowadays an essential prerequisite for many polymer studies. GPC has, over the past 15 years, become a well established procedure for the fractionation and precise characterization of high polymers. Although significant progress has been made in the understanding of the mechanism of GPC, a general and rigorous model for the separation process has still not been achieved. Many systems have been studied and empirical relations were obtained for the dependence of the elution volume on the molecular weight (MW), but most theoretical work was limited to very simplified pore network geometries and to idealized polymer shapes.

In the present study, both MW calibration and the effect of experimental variables on the shape of the elution peak were thoroughly investigated. Controlled pore glass (CPG) was used as separation matrix because of its unsurpassed mechanical stability, chemical inertness to the system under investigation and constancy of pore structure. Dextran was selected as a polymer because of its well known conformational and solution properties, and the availability of fractions of extremely narrow MWD and precisely known MWs.

EXPERIMENTAL

Dextran fractions with narrow MWDs ($\overline{M}_w/\overline{M}_n$ <1.06) were prepared from commercially available products by preparative GPC on Sephadex and Sepharose. The MW averages $\overline{M}_n$, $\overline{M}_w$, and $\overline{M}_z$ were determined by end-group analysis, low angle laser light scattering and sedimentation equilibrium. CPG columns having different pore diameters were calibrated with these dextran fractions as described previously[1].

Investigations were carried out on columns packed with CPG (particle size 75–125 μm) of different pore diameters in the range of 8.4 - 65.4 nm. Chromatographic runs were performed using standard equipment[2]. Eluant was water containing 0.2% KNO_3. For calibration purposes the coefficient $K = (V_e-V_o)/(V_t-V_o)$ was used; V_e is the volume at the maximum of the elution peak of the substance with MW M, V_o is the exclusion volume, and V_t is the total free volume of the column. To investigate the elution characteristics of the polymer fractions pulse sample elution and frontal experiments were carried out. The peak broadening function (PBF) was determined experimentally by using extremely narrow dextran standards ($\overline{M}_w/\overline{M}_n$ < 1.01); it was characterized by the standard deviation σ and the asymmetry A_s of the peak. A_s is defined by the ratio of the σ-values of the leading and the trailing branch of the peak.

RESULTS AND DISCUSSIONS

A typical set of calibration curves is shown in Figure 1 for CPG with different pore sizes. The calibration curves can be represented by straight lines if $lg(1-K)$ is plotted against $lg M$. From the straight lines thus obtained the critical MW of the polymer molecules having the same dimensions as the pore size of the considered separation matrix can be obtained by extrapolation of the MW to $K = 0$.

The elution volume V_e of dextran with a definite MW is almost independent of the experimental conditions under which the chromatograms were run. Even for slightly asymmetrical peaks, no dependence of V_e on the flow rate $\hat{V}$ is observed over a wide range, if V_e is taken at the bisectrix of the peak area (Fig. 2). These results indicate that equilibrium distribution of the polymer molecules is established between the mobile phase and the stationary phase within the pores.

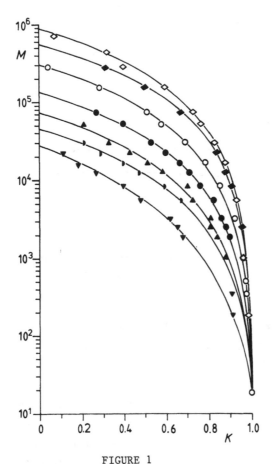

FIGURE 1

Pore Diameters: ◇ 65.4 nm; ◆ 51.7 nm; o 34.7 nm; ● 22.7 nm;
▲ 15.9 nm; ▶ 11.8 nm; ▼ 8.4 nm.

Investigations of peak broadening were carried out in three
steps: instrumental broadening (σ_i), broadening within the void
volume of the column (σ_v), and the combined broadening within the
void and the pore volume of the column (σ_c). σ_v was determined by
chromatography on nonporous glass; a direct relation between σ_v and
the diffusion coefficient of dextran was found.

Broadening within the pores of CPG is much more pronounced than
in the intersticial volume and increases with pore size. The influ-

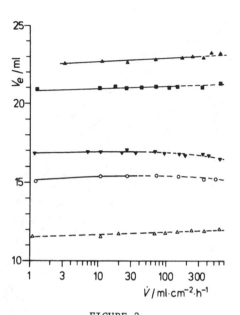

FIGURE 2

Δ glucose; ■ $\overline{M}_w$ = 11000; ▼ $\overline{M}_w$ = 90300; o $\overline{M}_w$ = 145000; Δ $\overline{M}_w$ = 5.2 x 10^6. Full Lines Refer to Gaussian Peaks, Dotted Line to Asymmetrical Peaks.

ence of the MW of dextran is comparatively small. Peak broadening depends, however, strongly on the flow rate $\dot{V}$. In Figure 3, the PBF is represented as a function of $\dot{V}$ for different dextran fractions (curves A); the PBF within the intersticial volume (curves B) and the instrumental broadening function (□) are included.

Peak asymmetry is also strongly dependent on the flow rate and increases for higher MWs (Fig. 4). The results of the present investigation indicate that peak asymmetry is caused mainly be the disturbed flow of eluant within the intersticial volume, around the irregularly shaped packing material.

Peak broadening was also investigated in frontal experiments. No differences in the corresponding values of the elution volume or peak shape were found between conventional elution and frontal experiments, irrespective of flow rate and MW of the polymer. This confirms that interactions between polymer molecules and the packing

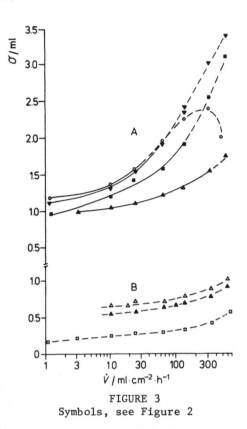

FIGURE 3
Symbols, see Figure 2

material of the column is negligible and insignificant for the mechanism of GPC.

The dependence of the separation parameters (V_e, σ, A_s) on flow
rate and packing characteristics of the column are discussed in connection with the phenomenological model of Ouano[3]. This model
treats the column as two continuous phases, a mobile phase in which
the molecules undergo molecular and eddy diffusion, and a stationary
phase where the molecules undergo partition and molecular diffusion.
The results of Ouano's simulations were confirmed in principle by
the experimental findings of the present study. The model, however,
gives no information on the coefficient K.

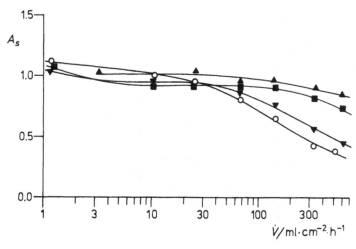

FIGURE 4
Symbols, see Figure 2.

        To calculate the coefficient K as a function of the MW of the
polymer and the pore size of the packing material, a stochastic
model[4] of the separation process was developed. It is assumed
that, as a polymer in dilute solution percolates through the column,
the molecules will achieve a partition equilibrium with distribution
coefficient K between the mobile phase and the stationary phase inside the pores of CPG. The polymer molecules are considered to be
random coils, showing no adsorption on the separation matrix, and
that conformational changes are not accompanied by enthalpic effects.
On these conditions it can be assumed that the free enthalpy changes
during the transference of a polymer molecule from the mobile phase
into the pores are caused exclusively by conformational entropy
changes due to the limited geometry of the pores. It can be shown
that the partition coefficient, K, usually defined as the ratio of
the concentration of polymer molecules within the porous medium to
the concentration in the mobile phase, is equal to the ratio of the
number of the allowed conformations of the polymer molecule within
the separation matrix to those in the mobile phase.

In the present study the number of allowed and forbidden conformations of a polymer with definite MW in a cylindrical pore of
radius r was calculated by means of a Monte Carlo method. As can be
verified by electron micrography, the pores of CPG are, in fact, of
almost cylindrical shape[5] as assumed in the simulations. More
than 2000 molecules were simulated for every MW. The universal
calibration curve was constructed by repeating this procedure for
about 50 MWs in the range from 180 to 10^6. In a double logarithmic
scale, a plot of (1-K) versus $M.r^{-2}$ gives an almost linear universal
calibration curve, which is represented by the solid line in Figure
5. The results of a large number of calibration runs on CPG of different pore size are inserted in Figure 5. The good agreement between simulation and experiment confirms the applicability of stochastic models for the elution behavior of polymers on columns of CPG.
The model developed in this study enables the calculation of calibration curves for random coils as a function of the pore radius of
CPG[6].

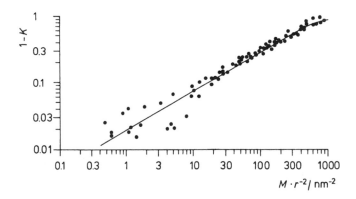

FIGURE 5
Solid Line: Stochastic Model. ●: Experimental Points.

REFERENCES

1. Basedow, A., Ebert, K.H., Ederer, H., Hunger, H., Makromol.
    Chemie, <u>177</u>, 1501 (1976).

2. Haller, W., Basedow, A., Konig, B., J. Chromatogr., <u>132</u>, 387
    (1977).

3. Ouano, A., Adv. Chromatogr., <u>15</u>, 233 (1977).

4. Casassa, E., J. Polym. Sci. Polym. Lett., <u>5</u>, 773 (1967).

5. Haller, W., Nature, <u>206</u>, 693 (1965).

6. Basedow, A., Ebert, K.H., Ederer, H., Fosshad, E., to be publish-
    ed.

POLYMER RETENTION MECHANISMS IN GPC ON ACTIVE GELS

J.E. Figueruelo and V. Soria　　　　　　A. Campos

Depto. Quimica Fisica　　　　　　Depto. Quimica Fisica
Facultad C. Quimacas　　　　　　Facultad de Ciencias
Universidad de Valencia　　　　Universidad de Bilbao
Burjasot (Valencia), Spain　　　　Bilbao, Spain

ABSTRACT

Elution behaviour of polystyrene (PS) and poly (methyl metha-
crylate) (PMM) on Shperosil in many different eluents and its dependence upon solvent goodness, as defined by the α exponent of the
Mark-Houwink eq., with solvent strength (ε^o), with flow rate and
with molecular size has been studied. The application of a network-
limited partition and adsorption mechanism allows the evaluation of
relative distribution coefficients (K_p) with reference to a standard
state, this being any good solvent eluent system.

Low K_p values are obtained, with PS as a solute, at high ε^o
eluent values, partition being mainly responsible. At low ε^o values,
adsorption effects contribute to the high K_p values besides partition. An increase in flow rate results on a decreasing in retention,
this effect being more evident the higher adsorption contributions
are. Similar trend is also observed when lowering the molecular
weight down to the oligomer region: partition not being apparently
affected by molecular size, while adsorption seems to diminish in
the lower molecular weight region.

In general PMM follows the above elution trend, its elution
curves being always shifted to higher retention volumes than those
of PS, probably due to adsorption effects.

INTRODUCTION

In gel permeation chromatography (GPC) on active supports, such
as with polar inorganic gels, the hydrodynamic volume curves for PS
in poor solvents shift in some cases to lower [1,2], in other to

49

higher [3,4] retention volumes in comparison with a good solvent.
Separation mechanisms in these gels will be highly influenced by
adsorption and retention phenomena due to the polarity of supports
(4,8).

     In this paper, those phenomena are deeply analyzed, extending
elution behaviour studies to the more polar poly (methylmethacry-
late), PMM, and to low molecular weight polymers. Changes in elution curves with flow rate, polymer nature and eluent as well as
with molecular size are explained in terms of polymer-gel, solvent-
gel and polymer-solvent interactions.

                            EXPERIMENTAL

     All measurements were carried out on a Waters Assoc. model
ALC/GPC 202 liquid chromatograph, equipped with a 6000 p.s.i. pump,
a differencial refractometer unit R401, a 440 absorbance detector
(λ_1 = 254 nm, λ_2 = 280 nm) and an U6K universal injector, admitting
sample sizes from 1 μl up to 2 ml. The differencial refractometer
cell was thermostated at 25°C and the elution volumes were determined by means of a Waters siphon system of 0.97 ml located at the
detector outlet. When working with solvent mixtures, the eluent
reservoir was connected to a vessel containing solvent mixture of
the same composition. The vessel was equipped with a cooling device which acts as a saturating agent. The constancy in mixture
composition was tested by refractometry. A Pharmacia SR 25 column
(45 x 2.5 cm I.D.) was used. Three different packings were employed.
Packing A consists a 74 g of Spherosil XOA 200 of high granularity.
Packing B is a mixture of 25 g of Spherosil XOA 200 and 25g of XOB
075 both of high granularity. Finally packing C, for low molecular
weight solutes, consists of a 25+25 g mixture of Spherosil XOA 400
+ Spherosil XOA 200. Along the paper these packings will be called
gel A, gel B and gel C respectively.

     Up to thirty-four PS samples with narrow molecular weight dis-
tributions, (M_w/M_n)<1,08 in all cases, covering the molecular weight

range from 450 up to 660000, have been used. Polymethyl methacry-
late samples are fractions of radical whole polymers, with polidis-
persity indices as determined by independent GPC measurements,
varying in the range 1.1-1.2.

Solvents were dried in the usual way and purified by succes-
sive distillations. Polymer solutions were prepared immediately
before injection, using solvent from the eluent reservoir.

Eluent volumes in pure solvents of samples with molecular
weights higher than 50,000 were determined by extrapolation to
zero concentration. This was not necessary in ternary solutions
near or at Θ composition[5] and in pure solvents for polymers with
molecular weights lower than 50,000. In these last cases a solution concentration of 3 mg/ml was used, the injection volume depending on the difference in refractive index between polymer and
eluent.

Intrinsic viscosities were measured with a conventional Ubbelhode viscosimeter and α values of the Mark-Houwink-Sakurada (MHS)
equation were calculated from the molecular weights determined by
GPC on μ-Styragel with THF as eluent.

RESULTS AND DISCUSSIONS

Previous detailed studies[4,8] on the elution behaviour of PS
on gel A in a great variety of eluent systems (twenty-one) at a
flow rate of 1.0 ml/min. allowed us to draw some useful conclusions.
Therefore, in the following, the most remarkable findings o those
studies, together with new data on gel B, will be covered first
under the heading polystyrene.

<u>Polystyrene</u>

Some of the overall calibration curves, $\log(M[\eta]$ vs. $V_R)$, obtained
with gel type A at a flow rate of 1.0 ml/min. are pictured in Figure
1. The total vlume of available pores for each eluent system is the
difference V_0-V_m, V_0 being the interstitial volume or "void volume"
and V_m the total permeation volume. V_0 values have been obtained
from the elution volumes of a PS of molecular weight 660,000 and

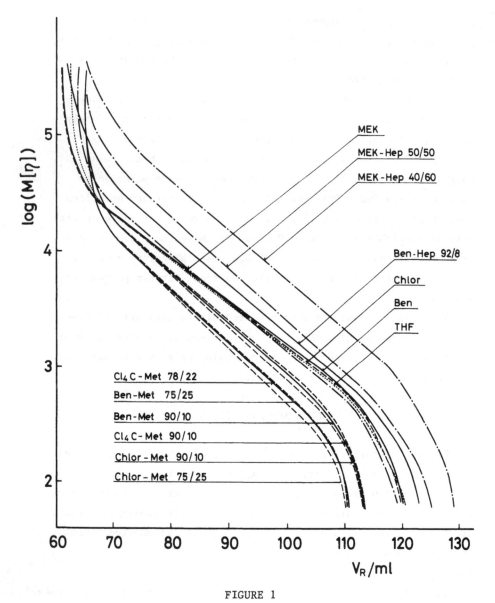

FIGURE 1

PS Elution Curves on Gel A in Different Eluents.

there is no doubt that on gel A the top limit for species separation
is reached with such a molecular weight. Thus, for example, in the
eluent showing the minimum hydrodynamic volume, namely chloroform-
methanol (75/25), the radius of the equivalent hydrodynamic sphere
of PS 660000 is 183 $\overset{\circ}{A}$. In these zones the polymer does not penetrate into the pores, but V_0 values are different for each eluent
system, being minimum (60.7 - 60.8 ml) for good solvents and maximum (65.0 - 65.2 ml) for mixtures near to or at Θ composition, as
it is illustrated by V_0 and α values collected in Table 1. V_0 values
seem to be only dependent on the thermodynamic quality of solvent
and to be independent of solvent strength (see also, in Table 1, ε°
(Al_2O_3) values) and even of the nature of support, since similar
trends have also been observed with crosslinked PS gels[9,10].

   V_m values have been taken as the elution volumes corresponding to PS of $\overline{M}_n$ = 2000. In Table 1 the size of this macromolecule
in different eluent systems expressed as the perturbed mean radius
of gyration, $<s^2>^{1/2}$, and as the equivalent hydrodynamic sphere
radius, R, is given. $<s^2>^{1/2}$ has been evaluated from

$$<s^2>^{3/2} = [\eta]M/ 6^{3/2}\Phi(\varepsilon) \tag{1}$$

where $\Phi(\varepsilon)$ is calculated by the classical Ptitsyn-Eizner equation
[11], with Φ_0 = 2.87 x 10^{21}. Hydrodynamic radii have been calculated from the Einstein equation

$$R^3 = 30 M[\eta] 10^{24}/\pi N_A \tag{2}$$

   As indicated above, the pore radii of gel range between 50 -
100 $\overset{\circ}{A}$ and according to the pore size distribution curve determined
by mercury porosimetry the volume of pores with radii smaller than
35 $\overset{\circ}{A}$ is inestimable, this last value being far from the R and $<s^2>^{1/2}$
values of Table 1. PS with $\overline{M}_n$ = 2000 is then on the limit of total
permeation and it will practically penetrate into every pore; its
elution volume, V_m, minus the respective V_0 will yield the total
available volume of pores in every solvent.

   Partition and/or adsorption effects must be invoked to explain
the large deviations in elution curves of Figure 1. In the network-
limited partition and network-limited adsorption mechanism, pro-

TABLE 1

Parameters Defining PS Elution Behaviour on Spherosil Type A in Different Eluent Systems.

Eluent	V_o/ml	V_m/ml	α (*)	$\varepsilon^o(Al_2O_3)$	$\sqrt{\langle s^2\rangle}$/Å	R/Å
Benzene	60.8	120.2	0.67	0.32	12.5	10.7
Benzene-Methanol (90/10)	65.1	113.3	0.60	0.72	12.3	10.6
Benzene-Methanol (84/16)	65.3	112.1	0.55	0.80	12.4	10.7
Benzene-Methanol(75/25)	65.2	110.6	0.50	0.89	12.1	10.4
Chloroform	60.7	119.8	0.68	0.40	12.5	10.6
Chloroform-Methanol (90/10)	65.2	113.1	0.61	0.71	12.1	10.3
Chloroform-Methanol (84/16)	64.9	111.8	0.56	0.78	12.0	10.4
Chloroform-Methanol (75/25)	65.2	110.3	0.50	0.87	11.9	10.4
Benzene-n,Heptane (92/8)	62.1	123.0	0.66	0.31	12.5	10.7
Benzene-n,Heptane (70/30)	----	----	0.56	0.27	----	----
2-Butanone (MEK)	63.4	119.3	0.56	0.51	12.4	10.6
MEK-n,Heptane (75/25)	63.5	121.0	0.57	0.45	12.5	10.6
MEK-n,Heptane (60/40)	64.6	123.4	0.54	0.41	12.1	10.5
MEK-n,Heptane (50/50)	64.8	125.5	0.51	0.39	12.0	10.5
MEK-n,Heptane (45/55)	65.0	127.7	0.50	0.37	11.9	10.5
MEK-n,Heptane (40/60)	65.1	129.4	0.50	0.36	11.9	10.5
Tetrahidrofuran (THF)	62.4	120.7	0.61	0.45	12.5	10.7
Carbon Tetrachloride (TC)	----	----	0.68	0.18	----	----
TC-Methanol (90/10)	65.0	113.4	0.61	0.74	12.4	10.7
TC-Methanol (85/15)	65.2	112.3	0.56	0.82	12.3	10.7
TC-Methanol (78/22)	65.1	110.5	0.50	0.88	12.2	10.5

(*) In the molecular weight range 16,000 - 80,000

posed by Dawkins and Hemmins[9,12], the equation relating retention
volume to distribution coefficient and average pore radius is

$$\ln(V_R - V_o) = -\frac{R}{\bar{r}} + \ln\frac{2K_p}{\bar{r}} \tag{3}$$

where V_R stands for elution volume. K_p is the distribution coefficient for solute partition between stationary and mobile phases
(if $K_p = 1$ there is no retention and solutes will separate by steric
exclusion alone). $\bar{r}$ is the average pore radius of the gel and, finally, V_0 and R stand, respectively, for void volume and hydrodynamic
radius of the macromolecules, as stated before. According to eq.
(3) a plot $\ln(V_R - V_0)$ vs R will yield a straight line, $\bar{r}$ being obtained from the slope. In Figure 2 such plots have been represented
for some eluent systems. The resulting $\bar{r}$ values for all the eluents
are given in Table 2. The radii are always smaller than the average
pore radius given by the manufacturer ($\bar{r} = 75$ Å) and they seem to be
dependent on eluent strength as Figure 3(a) shows. For $\varepsilon^o > 0.45$, in
some cases for slightly smaller ε^o, a linear dependence with negative slope between average pore radius and ε^o exists. When $\varepsilon^o < 0.45$,
$\bar{r}$ values are shifted to high values, the deviations increasing as
solvent strength decreases. This seems to indicate that now $\bar{r}$ does
not depend solely on solvent affinity for gel and some new factor
must cause such a displacement. Thus, when solvent-gel interactions
are sufficiently small, they will compete with polymer-gel ones,
which, in turn, will be enhanced when polymer-solvent interactions
decrease. The new factor representing polymer-solvent interactions
may be the α exponent of the MH equation. In Figure 3(b), $\bar{r}$ vs α
is plotted in the zone $\varepsilon^o < 0.45$, the expected dependence being observed. For small α values, the polymer, which interacts weakly
with solvent, will tend to interact with gel, making that V_R as
well as $\bar{r}$ increase. Therefore, in a first approximation, and in
those zones of solvent strength in which the existence of a compe-
tence between polymer and solvent for gel has experimentally been
detected, both ε^o and α measure the strength of polymer-gel inter-

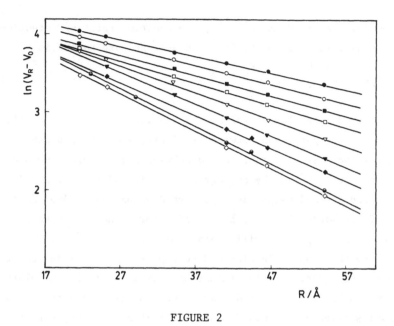

FIGURE 2

Eq. (3) Plots for Same Eluent Systems: (●) MEK - n-Heptane (40/60);
(O) MEK - n-Heptane (45/55); (■) MEK - n-Heptane (50/50); (□)
MEK - n-Heptane (60/40); (▽) MEK ; (▼) Carbon Tetrachloride -
Methanol (90/10); (◆) Carbon Tetrachloride - Methanol (85/15);
(◑) Carbon Tetrachloride - Methanol (78/22); (◇) Benzene - Metha-
nol (75/25).

actions, since the decrease of both separately influences the
increase of $\bar{r}$.

The joint dependence of $\bar{r}$ on ε^o and α, for $\varepsilon^o < 0.45$ is shown
in Figure 3(c) with two kinds of functionality, namely $\varepsilon^o\alpha$ and
$1/\varepsilon^o\alpha$. $\bar{r}$ values are also arranged in this zone, the limit of
experimental observation of $\bar{r}$ lying at $\varepsilon^o\alpha = 0.17$.

It is observed that $\bar{r}$ values decrease as ε^o increases and this
seems to indicate that the effective radius of the pores decreases[3]
due to the formation of a layer of "quasistationary" phase of interacting solvent which occupies the active sites in gel pores. The
volume of solvent layer covering the gel pores will be proportional

TABLE 2

Average Pore Radius and PS Relative Distribution Coefficients, f, on Spherosil XOA 200 in Different Eluents.

Eluent (Vol./Vol.)	$\bar{r}/Å$ (a)	I (b)	f
Carbon tetrachloride (TC)	no recovery	----	very large
Benzene-n,Heptane (70/30)	no recovery	----	large
MEK-n,Heptane (40/60)	46.0	4.56	1.49
MEK-n,Heptane (45/55)	40.0	4.54	1.28
MEK-n,Heptane (50/50)	39.3	4.42	1.11
Benzene-n,Heptane (92/8)	37.0	4.42	1.05
MEK-n,Heptane (60/40)	37.0	4.39	1.01
MEK-n,Heptane (75/25)	31.5	4.54	1.00
Reference System (c) — Chloroform, THF, Benzene, MEK	31.4	4.54	1.00
TC-Methanol (90/10)	24.0	4.63	0.84
Benzene-Methanol (90/10)	23.9	4.56	0.78
Chloroform-Methanol (90/10)	25.1	4.48	0.75
TC-Methanol (85/15)	22.5	4.59	0.75
Benzene-Methanol (84/16)	22.6	4.51	0.70
Chloroform-Methanol (84/16)	22.6	4.51	0.70
TC-Methanol (78/22)	21.2	4.52	0.70
Chloroform-Methanol (75/25)	20.7	4.61	0.71
Benzene-Methanol (75/25)	20.1	4.56	0.65

(a) Average pore radius from eq. (3)

(b) Fig. 2 intercepts

(c) $\bar{r}$ values for chloroform, THF, benzene and MEK are respectively 32.2, 31.0, 33.0 and 29.3 Å.

to the mean radius of pores free of solvent ($\bar{r}_f$ = 75 Å) minus the actual mean radius, $\bar{r}$, to the third power. A plot ($\bar{r}_f - \bar{r})^3$ vs. ε^0, as shown in Figure 4, will indicate the influence of solvent strength on the volume of solvent layer. Again, for ε^0 higher than the critical one, a straight line is followed. It goes through the origin as must be expected when the solvent layer disappears. Moreover, in systems with ε^0 and α small enough the volume of solvent layer becomes very small and deviations from linearity appear. This phenomenon occurs in the systems PS-butanone-n,heptane at haptane volume fractions equal or higher than 40%. However, in systems as PS-ben-

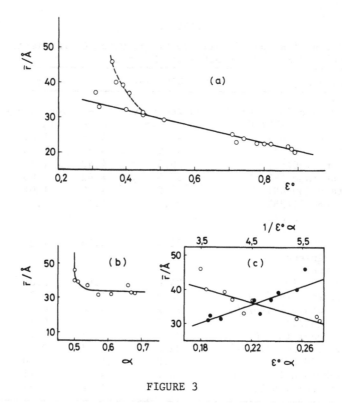

FIGURE 3

(a) Average Pore Radius Dependence on Solvent Strength. (b) Average Pore Radius Dependence on Solvent Goodness for Eluent Systems
with $\varepsilon^0 < 0.45$. (c) Joint Dependence on Solvent Strength and Solvent
Goodness of Average Pore Radius for Eluent Systems with $\varepsilon^0 < 0.45$.

zene-n-heptane (92/8) and PS-benzene, with respective ε^0 values of
0.31 and 0.32, lower than the ε^0 values of any butanone-n-heptane
mixtures, but with respective α values of 0.66 and 0.67 higher than
the α values of any of the butanone mixtures, the polymer is sufficiently fixed by the solvent as not to interact with gel, the volume of solvent layer in those systems still being linear with ε^0.

According to Figures 3 and 4 (a), the PS-butanone-n-heptane
(40/60) system with $\varepsilon^0 = 0.18$ is on the limit of polymer recovery.
Therefore, it is not surprising for the polymer in the PS-benzene-n-
heptane (70/30) system with $\varepsilon^0 = 0.15$ to be recovered with great dif-

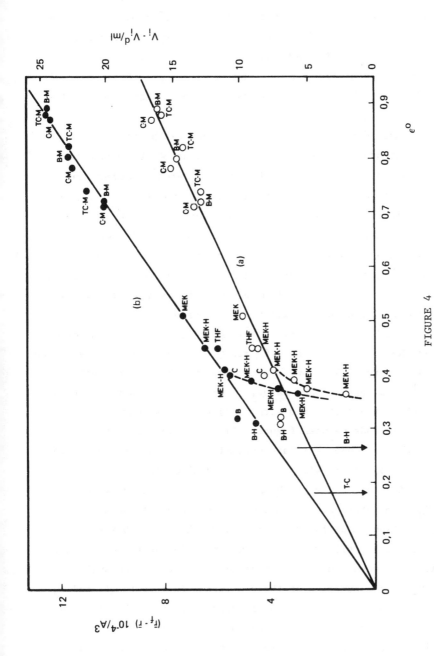

FIGURE 4

Dependence on Solvent Strength of (a) Average Volumes of Solvent Layer in Pores and (b) Total Volume of Solvent Layer. B=Benzene; C=Chloroform; H=n-Heptane; M=Methanol; MEK=2-Butanone; TC=Carbon Tetrachloride; THF=Tetrahydrofuran.

ficulty , without any possibility of establishing a calibration and
that in the PS-carbon tetrachloride system with $\varepsilon^0 = 0.12$, the polymer is not recovered at all. Eltekov and Nazansky's adsorption
measurements confirm this last behaviour: the amount of adsorbed
PS in carbon tetrachloride on macroporous silica gel (Silochrom C-2)
is about 1 mg/m^2, the thickness of the PS adsorbed layer being 10 -
15 Å [13].

The above ideas on solvent layer volume can be tested in a
purely experimental way, as shown also in figure 4 (b), where the
volume of solvent layer, expressed by the difference between the
total volume of pores as determined by mercury porosimetry, V_i, and
the accessible volume of pores for each eluent system, $V_i^d = V_m -
V_0$, have been represented versus ε^0. The Spherosil XOA 200 manu-
facturer gives an average pore diameter of 140 Å and a specific surface of 140 m^2 . g^{-1}. From pore size distribution curves an average
pore diameter of 150 Å and a volume of pores between 0.9 and 1.0 ml.
g^{-1} = 70.3 ml. 74 g being the weight of gel used.

The straight line $V_i - (V_m - V_0)$ vs. ε^0 goes through the origin
confirming the goodness of the 0.95 ml. g^{-1} value taken for the volume
of pores of Spherosil XOA 200. Pore volume values for macroporous
silicas[13] also confirm that value, in spite of the fact that this
gel has different diameter and specific surface than ours.

The similitude between figures 4(a) and 4 (b) saves any discussion on the last one, similar conclusions to those arrived to in
the discussion of figure 4(a) being applied also here.

According to Dawkins and Hemming [9], the quantitative evaluation of K_p from intercepts of Figure 2 is not possible because of
the approximations made in the derivation of Eq. (3). However, we
believe Eq. (3) is still valid to obtain K_p relative values. Thus,
the relation between the distribution coefficients of two systems
(K_{p1} and K_{p0}) as a function of their Figure 2 intercepts (I_1 and I_0)
and average pore radii ($\bar{r}_1$ and $\bar{r}_0$) is expressed by the Eq. (3). A
problem arises when intending to assign the numerical values of
those relations, as does the selection of the reference systems. The

intercepts of Figure 2 are so close for all the systems as to make
their differences small. Therefore, K_{p1}/K_{p0} values will mainly be
governed by $\bar{r}_1/\bar{r}_0$ ratios.

If it is taken as the reference system, e. g. benzene, THF,
... and the value $K_p = 1$ is assigned to it, higher, but also lower
than unity K_p values will be obtained. The problem, then, is to
give a physical meaning to the $K_p < 1$ values, since if $K_p = 1$ means
pure exclusion mechanism and $K_p > 1$ limited partition and/or adsorption mechanism, which mechanism is followed by $K_p < 1$ systems?

Therefore, in order to get K_p values higher than 1 for all
the other systems, the systems with smallest $\bar{r}$ should be selected
as the reference one, with $K_p = 1$. That system (see Table 2) is
the benzene-methanol (75/25) which, as must be expected, also
deviates the most towards low elution volumes as Figure 1 shows.
At this point, one must wonder if the benzene-methanol (75/25)
is the most appropriate reference system, because of Figures 3
and 4 seem to indicate and the above discussions on $\bar{r}$ and $V_i-V_i^d$
point to, systems may exist showing negative deviations with
respect to benzene-methanol (75/25), namely, those with $\alpha=0.50$
and $\varepsilon^0>0.89$, this last figure being the solvent strength value
of the benzene-methanol (75/25) eluent system.

The introduction of a coefficient $\underline{f}$ defined as

$$K_{p1} = \underline{f} \, K_{p0}$$

will overcome the above difficulties in the selection of an appropriate standard state, because, as already indicated[4], one
is generally interested in relating the changes in elution curves
to the corresponding changes in K_p values. Therefore, $\underline{f}$ will
be a coefficient showing the deviations of K_p value of system
1 with respect to that of the reference system 0 and it will be
given by

$$\underline{f} = \frac{\bar{r}_1}{\bar{r}_0} \, e^{I_1 - I_0} \tag{4}$$

Moreover, $\underline{f}$ will encompass the differences with respect to the
standard system in polymer-gel and solvent-gel interactions.

A convenient standard state would be one which is very often used. Of course, any polymer-good solvent system fulfills that condition. In our case there are four good solvent eluents, namely, benzene, chloroform, THF and butanone. Their elution curves in Figure 1 are very close and the same occurs with their $\bar{r}$ values (Table 2) and Figure 2 intercepts. On these grounds, the mean intercept and $\bar{r}$ values of the four systems have been respectively used as I_0 and $\bar{r}_0$ in Eq. (4). In Table 2, Figure 2 intercepts (1) and $\underline{f}$ values are shown.

Similar trends to those found on gel A are also followed on gel B. $\underline{f}$ values obtained on this last gel in new eluents are: in dioxane ($\varepsilon^0 0.56$, $\alpha=0.71$) $\underline{f} = 0.92$; in ethyl acetate ($\varepsilon^0=0.58$, $\alpha=0.71$) $f=0.86$. Many of the trends followed by $\bar{r}$ are also followed by $\underline{f}$. Therefore, in order not to make this paper unnecessarily long, the reader is submitted to the discussion on $\bar{r}$ values and their dependences with ε^0, α and $\varepsilon^0\alpha$. However, one point deserves some comments. As Determann pointed out [14], it is not easy to distinguish between partition and adsorption effects in GPC experiments. In our case, some clues enable us to venture some conclusions in that respect. These are the following:

(a) The strong relative sorption of methanol on Spherosil XOA 200 in benzene- and chloroform-methanol systems [4].

(b) The strong relative adsorption of PS on macroporous silicas in carbon tetrachloride [13].

(c) The variation in thickness of solvent layer in pores with ε^0.

For eluent systems with high methanol content and therefore high ε^0 and $\underline{f}<1$ values, the thick quasistationary layer of eluent strongly interacting with substrate prevents the polymer from approaching the gel. The solute will display its affinity for mobile and liquid stationary phases, which can differ in their composition [6]. The situation is similar to that displayed in liquid-liquid chromatography. Partition will be predominantly responsible for $\underline{f}$ changes.

On the other extreme, low ε^0 and $\underline{f}$>1 values, the layer
thickness decreases, allowing the polymer to approach the gel.
Besides partition effects, adsorption of solute onto the gel
starts to play a role (liquid-solid chromatography). At the lowest ε^0 (carbon tetrachloride) the PS relative adsorption on gel
is so strong as to produce an irreversible situation and the polymer
is not recovered at all.

In conclusion, at high ε^0 values partition rather than adsorption is the predominant effect governing $\underline{f}$ values. When ε^0 decreases, increasing adsorption effects must be added to partition ones.

Low Molecular Weight Polymers

Thermodynamic quality of solvents, as expressed by α expon-
ents, decays to the value α=0.50 with molecular weight of polymer.
Therefore, at molecular weights low enough, the sizes of macromolecules in solution becomes constant irrespective of solvent and
the task of separating steric exclusion primary contributions from
secondary partition and/or adsorption effects is eased. Figure 5
displays the elution behaviour of low molecular weight PS down to
the monomer in THF and benzene-based mixtures. Gel C, with smaller
pore diameter than gels A and B used previously, has been employed
as a support in order to enhance exclusion effects. In Figure 6,
butanone-based mixtures are the eluents.

Experimental results of Figures 5 and 6 may be summarized
as:

(a) Elution behaviour of low molecular weight PS is similar
in a qualitative way to that followed by high molecular weight PS,
as a comparison between Figure 1 and Figures 5 and 6 shows:

-Elution curves in THF and in benzene are very close.

-n-Heptane mixtures deviate to larger elution volumes than
the above standard eluents.

-Methanol mixtures, on the contrary, deviate to smaller
retention volumes.

(b) In mixtures deviating to the left, the magnitude of the
deviation increases with decreasing molecular weight, whereas in

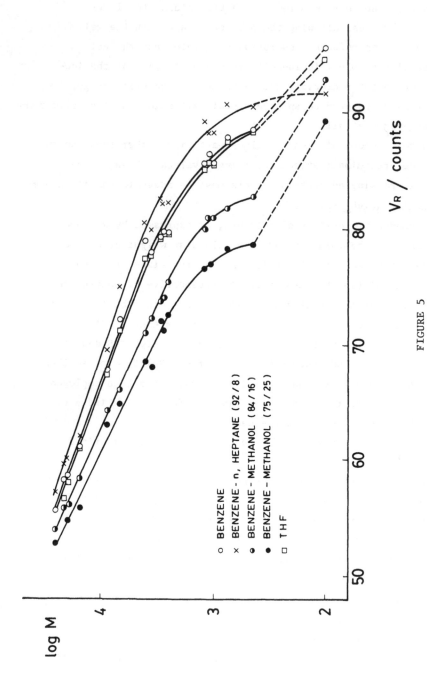

log M

V_R / counts

BENZENE
BENZENE - n, HEPTANE (92 / 8)
BENZENE - METHANOL (84/ 16)
BENZENE - METHANOL (75 / 25)
T H F

FIGURE 5

Elution Curves of Low Molecular Weight PS on Gel C in THF and in
Benzene and Benzene-Methanol Mixtures. 1 Count = 0.97 ml.

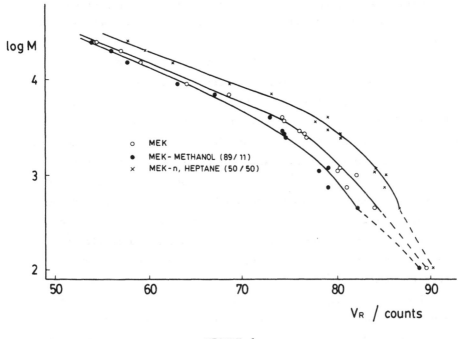

FIGURE 6

Elution Curves of Low Molecular Weight PS on Gel C in Butanone (MEK)
and in Butanone-Based Mixtures. 1 Count = 0.97 ml.

those going to the right it seems that the trend is only followed
at higher molecular weights; at lower molecular weights the differences in elution volumes remain constant (benzene - n-Heptane
mixture) or even decrease (butanone - n-Heptane mixture).

(c) Monomer elution volumes do not fit in any case to the
elution curves followed by polymer molecules, except for the benzene - n-Heptane (92/8) eluent. To illustrate this effect, dotted
lines have been drawn between the lowest molecular weight polymer
molecule ($\overline{M}_n$ = 450) and the monomer (the lowest points of Figures
5 and 6).

No mutual crossing of calibration curves in benzene and its
methanol mixtures takes place as suggested by Bakos et al[6]. In
the region of low molecular weights, the above results confirm

the conclusions arrived at in the study of higher molecular weight PS's. In those systems showing no polymer adsorption, namely methanol mixtures with $f<1.0$, the lower is the molecular weight, the smaller are the differences in the elution volumes with respect to the reference system, as must be expected. On the other hand, in those systems with $f>1.0$, in which adsorption starts playing a role, f values may be shifted towards the unity with decreasing molecular weight, since numerous studies indicate[15] that the adsorption of polymer increases with molecular weight and, as Dubin, et al. have pointed out [7], site interactions between polymer and substrate can exhibit cooperative neighbouring effects increasing with molecular weight. This may explain the apparent shortening of differences in elution volumes of n-Heptane mixtures with respect to the reference systems occuring in the lowest molecular weight region.

Elution behaviour of monomer is also in accordance with the scheme so far developed. The unsaturated character of monomer must increase its affinity for surface silanol groups against that displayed by polymer. As a result, higher retention of the monomer must be expected and Figures 5 and 6 evidence that in all the eluents. Caution is then required when intending to determine total permeation volumes (V_m) from monomer elution volumes, as it is done as a normal practice in GPC work.

Flow Rate and Other Polymers

Elution curves of PMM ranging in molecular weight from 10400 to 64000 and of PS just for comparison (in the molecular weight range 11000 to 60000) on gel B at elution rates of 0.6, 1.0, 1.5 and 2.0 ml/min have been determined in chloroform, chloroform-methanol (90/10), THF, butanone, ethyl acetate and dioxane. The following points briefly describe the results obtained:

(a) PMM eluates later than PS in all the eluents, qualitative as well as quantitative differences in elution curves being dependent on the eluent. In chloroform-methanol (90/10), THF, dioxane and butanone the elution behaviour of PMM may be considered as "normal" with respect to that of PS, deviations in elution curves

monotonically increasing when molecular weight decreases. This
"normal" behaviour in butanone and in chloroform-methanol (90/10)
is represented in Figures 7 and 8.

    (b) Figures 8 and 9 respectively show the "abnormal" PMM
elutions in chloroform-methanol (90/10) is represented in Figures
7 and 8.

    (b) Figures 8 and 9 respectively show the "abnormal" PMM
elutions in chloroform and ethyl acetate.

    (c) Both PMM and PS elution curves displace to lower re-
tention volumes on increasing flow rate, the magnitude of displacements being larger when PMM is the solute. This effect is
clearly observed in Figures 7, 8 and 9.

    In this paper, the importance of polymer-gel interactions as
opposite to solvent-gel/polymer-solvent interactions in governing
elution volumes on inorganic carriers has been stressed. The

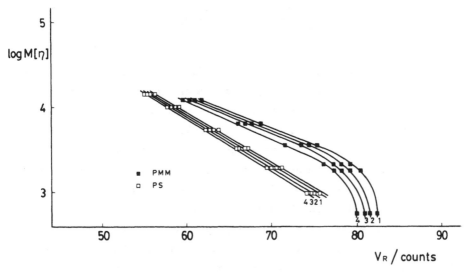

FIGURE 7

PS and PMM Elution Curves at Different Flow Rates on Gel B in Butanone. Flow Rate is Defined by: 1 = 0.6 ml/min; 2 = 1.0 ml/min;
3 = 1.5 ml/min; 4 = 2.0 ml/min. 1 Count = 0.97 ml.

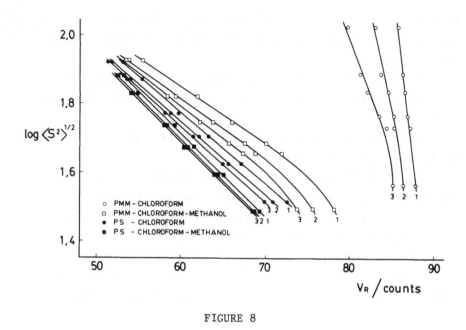

FIGURE 8

PS and PMM Elution Curves at Different Flow Rates on Gel B in Chloroform and in Chloroform-Methanol (90/10). Flow Rates are Defined by:
1 = 1.0 ml/min; 2 = 1.5 ml/min. 1 Count = 0.97 ml.

shift to higher retention volumes of PMM with respect to PS is a
further example. Stronger PMM polarity makes to increase sub-
strate-polymer interactions and therefore to increase its retention volumes. Chloroform is the only eluent lacking hydroxyl and
even the exact nature of polymer-substrate interactions are not
clear; the "abnormal" behaviour of PMM in chloroform seems to indicate the existence of strong interactions between carboxyl groups
of polymer molecules and the gel silinol sites. Probably, those
interactions are similar to thos displayed by small molecules
(ketones, alcohols and acids), which adsorb on the silanol groups
via hydrogen-bonding[16]. The addition of hydroxyl groups to
chloroform under the form of methanol normalizes the PMM elution
curves as figure 8 demonstrates. An experimental point must be
indicated before discussing the abnormality of PMM curves in ethyl

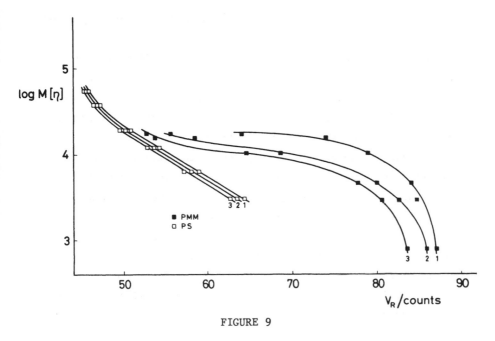

FIGURE 9

PS and PMM Elution Curves at Different Flow Rates on Gel B in Ethyl
Acetate. Flow Rates are Defined by: 1 = 1.0 ml/min; 2 = 1.5 ml/min;
3 = 2.0 ml/min. 1 count = 0.97 ml.

acetate. It is the difficulty of getting reproducible elution
volumes. Points appearing in Figure 9 are the mean values of four
to six determinations, deviations between them being as large as
2 - 3 counts. The similarity in chemical nature between eluent
and eluate is probably responsible for those difficulties and also
for the abnormal elution curves, substrate not being able to distinguish between eluent molecules and structural units of polymer molecules, cooperative effects in these last ones being important.

Negligible to slight differences in the peak maxima volumes
with flow rate both from theoretical [17] and experimental [18-20]
points of view, as well as distortions in peak shapes [17-21] have
so far been reported. Of course, those situations must be applied
to those systems in which exclusion by molecular size is the pre-

dominant mechanism, not being valid as evidenced by Figures 7, 8 and 9, when nonexclusion effects play a role.

Our lowest flow rates are in the range where diffusion into and out of the stationary phase is not a significant factor in the change of elution volumes, an equilibrium between stationary phase and solute existing[21], therefore, partition effects probably not being responsible for the observed differences in elution volumes with flow rate. However, the adsorption of the eluate on the gel surface needs previous desorption of enough solvent molecules[22] prior to the new competitive dynamic equilibrium between eluate and solvent molecules for the silanol sites is reestablished. It looks like if with increasing elution rate (minor residence time of eluate in the column) no time is let to the new equilibrium to be reached. Therefore, the stronger are eluate adsorption effects the higher are displacements in elution volumes with flow rate, as Figures 7, 8 and 9 seem to indicate.

ACKNOWLDEGEMENTS

Authors are grateful to the Comité Conjunto Hispano Norte-americano para la Cooperación Científica y Tecnológica (Project IIP 3080) for financial support of part of this work. Thanks are also given to Francisco I. Santiago (Bilbao University) for some elution measurements on gel B, to Miguel García (Bilbao University) for viscosity measurements and to Bernardo Celda and Vicente Sanz (Valencia University) for the reading of the manuscript.

REFERENCES

1. Swenson, H. A., Kaustinen, H. M., Almin, K. E., J. Polym. Sci., Part B 9, 261 (1971).

2. Berek, D., Bakos, D., Soltes, L., Bleha, T., J. Polym. Sci., Polym. Lett. Ed. 12, 277 (1974).

3. Berek, D., Bakos, D., Bleha, T., Soltes, L., Makromol. Chem. 176, 391 (1975).

4. Campos, A., Figueruelo, J. E., ibid. 178, 3249 (1977).

5. Bleha, T., Bakos, D., Berek, D., Polymer 18, 897 (1977).

6. Bakos, D., Bleha, T., Ozina, A., Berek, D., J. Appl. Polym.
    Sci. 23, 2233 (1979).

7. Dubin, P. L., Koontz, S., Wright, K. L., III; J. Polym. Sci.,
    Polym. Chem. Ed. 15, 2047 (1977).

8. Campos, A., Soria, V. and Figueruelo, J. E., Makromol. Chem.
    (in press).

9. Dawkins, J. V., Hemming, M., ibid. 176, 1795 (1975).

10. Dawkins, J. V., Hemming, M., ibid. 176, 1815 (1975).

11. Ptitsyn, O. B., Eizner, Yu E., Soviet Phys.-Tech. Phys. (Eng-
    lish Translation) 4, 1020 (1960).

12. Dawkins, J. V., J. Polym. Sci., Polym. Phys. Ed. 14, 569 (1976).

13. Eltekov, Yu. A., Nazansky, A. S., J. Chromatogr. 116, 99 (1976).

14. Determan, H., "Gel Chromatography", Springer-Verlag, Berlin-
    Heidelberg-New York, 1968.

15. Lipatov, Yu. S. and Sergeeva, L. M., "Adsorption of Polymers",
    Halsted Press, New York, 1974.

16. Marshall, K., Rochester, C. H., Faraday Disc. 59, 117 (1975).

17. Ouano, A. C. and Barker, J. A., Sep Scie. 8, 673 (1973).

18. Meyerhoff, G., J. Polym. Sci. C, 21, 31 (1968).

19. Boni, K. A., Sliemers, F. A., Appl. Polym. Symposia 8, 65 (1969).

20. Campos, A., Borque, L., Figueruelo, An. Quim. 74, 701 (1978).

21. Glajch, J. L., Warren, D. C., Kaiser, M. A., Rogers, L. B.,
    Anal. Chem. 50, 1962 (1978).

22. Lloyd, L. and Snyder, R. in J. J. Kirkland "Modern Practice of
    Liquid Chromatography", John Wiley & Sons, New York, 1971.

PROBLEMS IN MULTIPLE DETECTION OF GPC ELUENTS

R. Bressau

BASF Aktiengesellschaft, Kunststofflaboratorium
Ludwigshafen am Rhein, West Germany

ABSTRACT

Many authors have frequently reported in the literature on the investigation of GPC eluents with several detectors working according to different measuring principles. This multiple detection enables information concerning the chemical heterogeneity, the differences in long chain branching when molecular size detectors are employed and allows the measurement of the absolute molecular weight range without previous calibration.

As in most cases, the multiple detection is effected in cells being spaciously separated, thus the dead volume of the connecting capillaries has to be taken into consideration in the evaluation. In spite of the exact determination of this geometric dead volume the interpretation of the measurements leads to unsatisfactory results. The cause of these disturbances is that on transferring the eluents from one cell to another, changes in the peak shape occur in addition to the well known shifting by the dead volume. In the present paper this effect has been demonstrated by means of experimental examples. It has been tried by means of some models to describe these effects theoretically and to give instructions to optimize the real experimental conditions.

INTRODUCTION

The use of multiple detectors for analysis of GPC eluents gives additional information about composition and structure of polymer samples[1,2]. It is also possible by using molecular-size detectors[3,4] to calibrate the columns without the use of narrowly distributed standards. In most of the cases the different detection methods are not combined in a single cell; between cells rather long connecting capillaries are usually present. The interpretation of multiple detection elution curves[1] is a point-to-point comparison and depends on the dead volume of the apparatus. Thus, the signal of one detector has to be shifted by the volume of the connecting tube.

In this paper it will be shown that the connecting tube disturbs not only by shifting the elution curves, but also by altering its shape. The results of the evaluation are, however, noticeably disturbed. There will be described some model calculations to explain these effects.

To demonstrate these effects we analyzed a few polystyrene samples using a GPC unit with a differential refractometer and a UV-absorption detector. Since the samples were homopolymers with a constant composition, the quotient of signals $c_{DR}/_{UV}$ of both the detectors according to Cantow[2] should be a constant for all elution volumes: the plot of this quotient vs. molecular weight should be parallel to the abscissa. In practice, however, the result is usually a more or less bent curve. In the example (Fig. 1, narrowly distributed polystyrene = Press. Chem. Corp., M_w = 110,000) both the detectors are connected together with their inlets and outlets. The dead volume was measured by the elution difference of the peak maxima of ethylbenzene (ΔV = 0.46 $\pm$ 0.01 count, 1 count ca. 0.3 ml); this value is much higher than the geometric dead volume. By variation of the ΔV-value, it is possible to optimize the quotient curve but in no case the curve is a parallel line to the abscissa. The most optimal value obtained for different GPC-analyses was 0.04 $\pm$ 0.01 counts. For identical values of $\overline{M}_w$ from the refractometer curve and the UV-curve a different ΔV-value of 0.53 $\pm$ 0.01 counts was neccessary. Furthermore, the ΔV-value is different for broad distribution curves: on the same GPC-unit we need ΔV=0.52 count for an optimal quotient curve and ΔV=0.43 count for the identical value of both

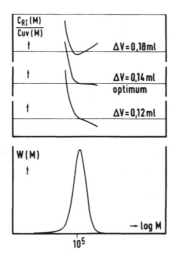

FIGURE 1
Multiple Detection of a Narrowly Distributed Polystyrene (=PCC 110000):
Influence of the Dead Volume Correction on the Quotient Curve C_{RI}/C_{UV}
According to Cantow. GPC Combination: 4 columns à 6 x 300 mm, Flow
Rate: 1.5 ml/min, THF Elution Volume at Peak Maximum: 22.4 ml.
V = .12 ml: Difference of Peak Maxima
V = .14 ml: Best Quotient Curve
V = .16 ml: Peak Difference of Ethylbenzene
V = .18 ml: Optimal Correcting Value for a Polystyrene Sample with
            a Broad Distribution, Analyzed on the same GPC unit.

the $\overline{M}_w$ values (the sample was a broadly distributed, radically ini-
tiated polystyrene).

      In all the experiments we found a greater apparent dead volume
than the geometric one. For an example of the different connecting
conditions see Table 6.

      The deforming effects are more easily eliminated after a bigger
column than after a smaller one. For example, the optimized quotient
curves for a polystyrene mixture analyzed by a 4.5 x 1200 mm and by
a 100 x 1200 mm column combination, respectively, are shown in Figure
10.

      These and similar observations show that the elution curves
are noticeably disturbed in their shape during the transition between the detector cells in a multiple detection system. We have

tried to simulate these shape deviations by model calculations. The
following models have been employed (see Fig. 2, 3).

    1. A radial distribution of the flow rate $u(r)$ according to
the law of Hagen-Poisseuille

$$u(r) = (R^2 - r^2) \cdot \Delta p / (L.4\eta)$$

R = radius of tube

Δp = pressure difference in a tube of length L

L = length of tube

η = dynamic viscosity

    2. Model #1 combined with an additional dead volume with
"perfect mixing". The results of our calculations are the
subject of this paper.

The validity of the law of Hagen-Poisseuille in a capillary under HP-GPC-conditions is justified by an estimation of the Reynolds
number:

$$R_e = \frac{v \cdot d \cdot \rho}{\eta} = 257.7$$

v = mean linear flow rate (for 1.5 ml/min = 60.94
cm/sec)

d = diameter of the capillary (here = 0.02248 cm =
.009 inch)

ρ = density, for THF = .888 g.cm

η = dynamic viscosity, for THF = 0.0048 $g.sec^{-1}.cm^{-1}$

The resulting value of 258 is distinctly lower than the critical Reynolds number of 1160. Therefore, we can safely assume a
laminar flow with a parabolical flow rate distribution in the connecting capillaries.

DESCRIPTION OF THE MODELS

Model #1 (see Fig. 2): A concentration $c_0(t)$ homogeneous over the
radius is present at the end fitting of a GPC-column combination.
The eluents flow through capillary #1 (length = L1) into the first
detector cell and then through the capillary #2 (length = L2) into
the second cell. It is assumed that the volume of the detector cell
is zero and that the flow profile is not disturbed in the cell. For
the calculation of the concentrations $c_1(t)$ and $c_2(t)$ in the cells

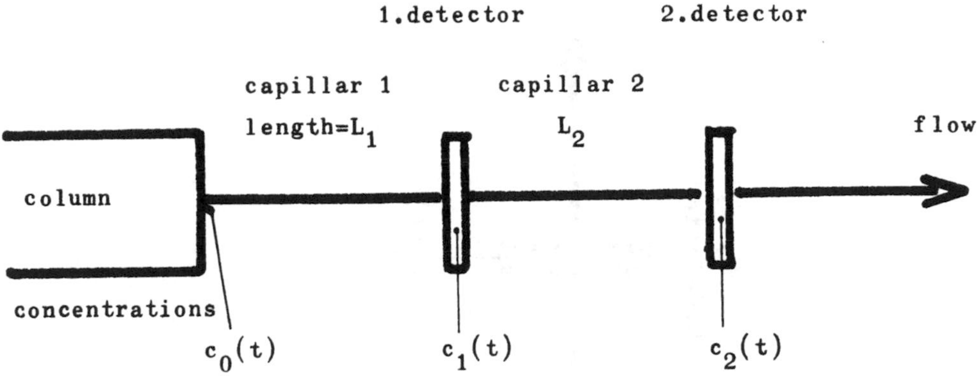

1.detector   2.detector

capillar 1   capillar 2

length=L_1 L_2 flow

column

concentrations

$c_0(t)$ $c_1(t)$ $c_2(t)$

FIGURE 2

Schematic of Model 1.

the concentration distribution is integrated on this point with respect to the radius and the mean value is taken.

Model #2 (see Fig. 3):   To model #1, a mixing chamber is added between the detector cells.   It possesses the properties of a "continuous stirred tank reactor", perfect mixing is assumed (in German:
"ideal durchmischter Durchlaufruehrkessel"[5]). The behavior of
this element is described by the following differential equation:

$$\frac{dc_{mix}}{dt} = \frac{Q}{V_{mix}} \cdot c_3(t) - \frac{Q}{V_{mix}} \cdot c_{mix}(t)$$

Q   = mean flow rate (Volume/time)

V   = volume of mixing chamber

c_3 = input concentration

c_{mix} = concentration in the chamber and at the outlet

Solution of the differential equation[6]:

$$c_{mix} = \exp[-\frac{Q}{V_{mix}} \cdot (t-t_0)] \cdot \left\{ \frac{Q}{V_{mix}} \cdot \int_{t_0}^{t} c_3(t) \cdot \exp[+\frac{Q}{V_{mix}} \cdot (t-t_0)] \, dt + c_{mix}(t_0) \right\}$$

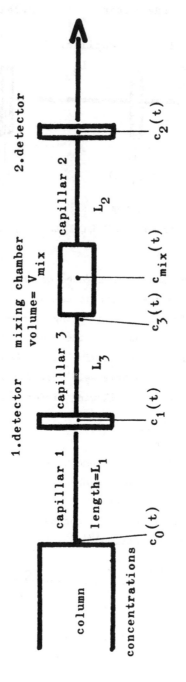

FIGURE 3
Schematic of Model 2.

This model has a series of special cases:

    a) $L3 = 0$, $L2 \neq 0$: mixing in the first cell immediately be-
       hind the sensitive area of the detector

    b) $L3$ and $L2 \neq 0$: mixing in a "poor" capillary union with
       high dead volume

    c) $L3$ and $L2 = 0$: "dualdetector": both the measuring princi-
       ples in one cuvette with a deadvolume with perfect mixing
       between both the sensitive areas

    For testing the properties of both models we are "eluating"
from the column a concentration profile like a normal (Gaussian) distribution: namely:

$$c_0(t) = \exp\,[t - \mu - \Delta t)^2 / 2 . \sigma^2]$$

μ and Δ are time correcting values for shifting the curves in the
first (=positive) quadrant and for producing a reference time for
all points of the model.

    After passing through a capillary of the length L the average
concentration across the diameter is changed from $c_0(t)$ to $c_L(t)$
according:

$$cL(t) = \frac{1}{R^2\pi} . \int_{r=0}^{r=R} 2r\pi . \exp[-(t - \mu - \Delta t - L/u(r))^2 / 2 . \sigma^2]\,dt$$

with $u(r)$ as the distribution of the flow rates over the cross-section of the capillary according to Hagen-Poisseuille.

    In the mixing volume of model #2, the distribution of the con-
centration over the cross-section is again made homogeneous. Therefore, $c_{mix}(t)$ is inserted as the starting concentration profile in
the formula for c_2:

$$c_2(t) = \frac{1}{R^2\pi} . \int_{r=0}^{r=R} 2r\pi . c_{mix}\,(t - L_2/u(r))\,dt$$

    The infinitesimal short mixing time of the continuous stirred
tank reactor has the consequence, that the volume influences the
mixing only by the ratio flow rate/volume but not as an additional
term in the sum of the dead volume L2 + L3. The assumption of laminar flow according to Hagen-Poisseuille has the consequence, that

the model #1 without a mixing volume is independent of the overall
flow rate. Further, the calculations show that the size of the de-
forming effects are only a function of the volume of the connecting
capillary: the results are independent of the radius of the capillaries so long as the connecting volume is kept constant.

The integrals were numerically computed according to the Simpson parabolic rule with step values of R/250 to R/600, the step
value for the mixing chamber was kept constant at 10% of V_{mix} . A
single integration step often results in a very low contribution to
the integral. Therefore, it was necessary to choose a double precision mode for the computation (i.e., representation of a floating-
point number by 64 bits).

Always the following characteristic values were being calculated:

   a) concentration $c_1(t)$ and $c_2(t)$ in both the detector cells
   b) the time and concentration of the peak maxima
   c) an asymmetry quotient A of an elution curve with (see
      Fig. 4)

$$\text{asymmetry } A = \frac{W_2}{W_1}$$

      In most of the cases the asymmetry A was lower than 1;
      that means the flow through a capillary lead to a back-
      tailing.

   d) analogue P of the plate number: in a similar method to
      the determination of the plate number we calculated the
      intersection points of the interflectional tangents with
      the abscissa and then the width W between both the inter-
      section points (see Fig. 4). For a non-disturbed normal
      distribution this value is exactly 4.sigma; therefore, we
      obtain a value of 1 for P in the case of a non-disturbed
      elution curve:

$$\text{analogue of the plate number } P = \left(\frac{1}{W/4.o} \right)^2$$

RESULTS AND DISCUSSION

Standard conditions

We are assuming as standard the following conditions:

a) the shortest possible connection between the end fitting
   of the last column and the 1st detector cell is a capillary of 25 cm length.

b) the shortest length between the 1st and the 2nd cuvette is
   50 cm even when the two detector units are directly connected by the original inlets and outlets.

c) the radius of the capillary is equal the lowest commercial
   available type with a diameter = 0.009' = 0.2286 mm.

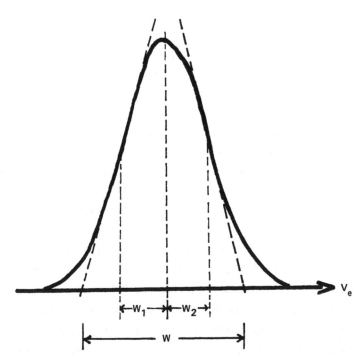

FIGURE 4

Elution Curve Illustrating the Band Width w and the Difference w_1 and
w_2 Between the Abscissa of Inflection Points and Peak Maximum used in
the Definitions of the Asymmetry A and of the Analogon P of Plate
Number.

    d) the 6-sigma-volume of the elution curve of a commercial
       narrowly distributed polystyrene calibration standard is
       about 2.0 ml on using a column combination of 4 columns of
       4.5 x 300 mm.

With these assumptions we obtain the following results by using
model #1 (see Table 1, Row 3):

    - The apparent dead volume between both the cells is about 50%
      greater than the theoretical (= geometric) one.

    - The distribution-width of the concentration curve is in
      cell #2 broader than in cell #1 (decrease of plate number P
      by about 3%).

    - The asymmetry A is in both the detector cells is slightly
      less than 1 and the backtailing is increased by the flow
      through the connecting capilliary #2.

These variations are seemingly low values but the effects on
the interpretation of the GPC-results are important: the point-
by-point quotient of the detector signals according to Cantow[2]
should be constant, equal to 1 and independent of the elution time
then both the detectors have the same sensitivity towards concentration. Assuming the geometric connecting volume as the dead volume
in the calculations results in a bent curve without a range parallel
to the abscissa (see Fig. 5). Shifting one of the elution curves by
the apparent dead volume (= calculated difference between the peak
maxima) gives a curve with at least a small range being approximately parallel to the abscissa (see Fig. 5).

    Further calculations show a decrease of disturbing effects with
a broadening of the elution curve or more generally formulated with
an increase of the 6-sigma-volume of the "eluated" concentration
profile (Fig. 5). That means that a polymer sample with a narrow
molecular weight distribution is more sensitive to distortion in a
multiple detector unit than a product with a broad distribution.
This is in agreement with the experimental observations. In practice it is possible to extend the elution range of a given sample by
increasing the diameter of the column. With constant length of the

TABLE 1

Model #1: Standard Conditions and Variation of the 6-Sigma-volume of the Eluated Curve; Flow Rate: 1.5 ml/min; Radius of the Capillaries: .01143 cm; Capillary #1: L1 = 25 cm; Capillary #2: L2 = 50 cm.

6-Sigma-Volume	Apparent Dead Time, Sec.(*)	Apparent-Geometric Dead Time, Sec.	Asymmetry A		Increase in 2.cell %	Analogue of Plate Number		Decrease %	Decrease of Max. Peak Height i.2. cell (%)
			1.cell	2.cell		1.cell P1	2.cell P2		
.9	1.00	.18	.994	.983	1.05	.961	.906	5.73	5.46
2.03	1.23	.41	.997	.992	.55	.981	.951	3.05	2.84
2.25	1.26	.44	.998	.992	.55	.983	.955	2.83	2.61
2.5	1.29	.47	.998	.993	.53	.985	.959	2.64	2.38
3.6	1.41	.59	.998	.995	.31	.989	.970	1.90	1.73
6.4	1.61	.79	.999	.997	.21	.994	.982	1.22	1.04
10.0	1.78	.97	.999	.998	.06	.995	.988	.77	.67
27.7	2.06	1.25	.9999	.998	.05	.998	.995	.28	.31

(*) = $t_{max2} - t_{max1}$

Approximated relations between the 6-sigma-elution volumes and the column radius (120-cm-combination of High-Performance-GPC-columns)

	3mm	4.5mm	6mm	8mm	10mm
6-sigma-volume of:					
narrowly distributed polystyrene (e.g. PCC 110000)	.9	2.03	3.6	6.4	10
broadly distributed polystyrene (radically initiated)	2.5	5.6	10	17.7	27.7

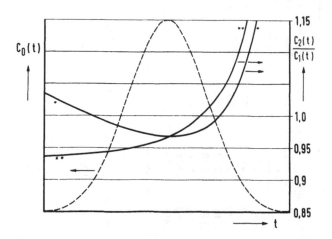

FIGURE 5
Results of Model #1: Standard Conditions. (s. text): Comparison
of the Quotient Curves $C_2(t)/C_1(t)$. Dead Volume Correction:
*) $C_1(t)$ Shifted by the Calculated Apparent Dead Volume
**) $C_1(t)$ Shifted by the Exact Geometric Dead Volume
--- = $C_0(t)$ = Concentration at the Column End Fitting.

column combination the 6-sigma-volume of the elution curve is proportional to R^2. In a series of computations we varied the 6-sigma-
elution volume of the concentration profile; the effects on the quotient curves $c_2(t)/c_1(t)$ are demonstrated in Figures 6 and 7. The
use of the exact geometric volume as the shifting value results in
no case in the theoretically expected straight line parallel to the
abscissa (Fig. 6). Only by correcting by the apparent time difference do we obtain quotient curves with parellel ranges (see Fig.
7). A reasonably broad range with a straight line approximately
parallel to the abscissa results only with the very large column
diameter of 1 cm.

To recapitulate: the necessary value for the best correction
is a function of the gradient of the detected property (e.g., c,
c.M, c.η) and cannot be determined by measuring either the geometric
dead volume of the connecting tube nor by the elution difference of
a homogeneous low molecular weight sample. Further for a given
multiple detector system an elution curve with a low concentration
gradient is less disturbed than a curve with a high gradient.

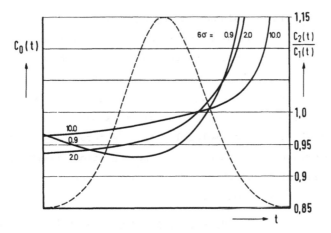

FIGURE 6

Quotient Curves $C_2(t)/C_1(t)$; Standard Conditions of Model #1 with Variation of the 6-sigma-volume of the Eluated Concentration Distribution. Correction of the Dead Volume by the Exact Geometric Dead Volume. --- = Concentration Curve $C_0(t)$ at the Column End Fitting.

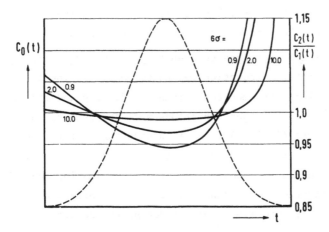

FIGURE 7

Quotient Curves $C_2(t)/C_1(t)$; Standard Conditions of Model #1 with Variation of the 6-sigma-volume of the Eluated Concentration Distribution. Correction of the Dead Volume by the Calculated Apparent Dead Volume. --- = Concentration curve at the Column End Fitting = $C_0(t)$.

Preferably large columns having the lowest concentration gradient during the elution of polymer samples should be employed. The
correcting value for elution curves possessing different sigmas
have to be determined by calibration using a series of homopolymer
samples of different molecular weight distributions. But this way-
out cannot be applied to a GPC-analysis of samples with molecular
weight distribution using molecular size detectors having response
curves proportional to c.M or c.η. These elution curves always exhibit a different shape to the corresponding curve of a concentration detector (proportional to c).

Variation of the connecting capillary between both the detectors

That the error increases on lengthening the connecting tube
is not surprising (Table 2); again a low concentration gradient is
less affected than a higher one. The size of the effect is remarkable: the elongation of the connecting capillary by only a length
of 50 cm decreases the plate number P and the maximal peak height by
about 5% (all other conditions equal to standard conditions). For
the effect on the quotient $c_2(t)/c_1(t)$, see Figure 8.

Variation of the capillary between the column and the 1st cell

In a further computation we changed only the length of the connecting capillary between the column end fitting and the cell of the
first detector, the other conditions were the same as in the standard conditions (Table 3). It is interesting that the geometry of
the previous capillary influences the shape deformation in a following connecting tube, however the effects are smaller and, therefore,
only of theoretical importance (e.g., peak deformation by a parabolic concentration profile developed during the separation in the
column).

Influence of the capillary radius

Model calculations confirm the long-known experience that the
disturbance increases with increasing capillary radius. This results in a lowering of the concentration curves, a broadening of
the peaks and an increase of the asymmetry A (Table 4). Surprising

TABLE 2

Model #1: Variation of the Capillary Length L2 between the Detector Cells. Flow rate: 1.5 ml/min; Radius = 0.1143cm; Capillary #1: L1 = 25cm; 6-Sigma-volume of Eluated Distribution = 2.25 ml.

Capillary Length L2		Apparent Dead Time Sec. (*)	Apparent -Geometric Dead Time Sec.	Asymmetry A			Analogue of Plate Number			Decrease of Max. Peak Height in 2. Cell(%)
cm	sec.			1.cell	2.cell	Increase in 2.cell	1.cell P1	2.cell P2	Decrease %	
50	.82	1.26	.44	.998	.992	.55	.983	.955	2.83	2.6
75	1.23	1.80	.57	.998	.990	.76	.983	.943	4.06	3.8
100	1.64	2.31	.67	.998	.988	.97	.983	.932	5.20	4.9

(*) = $t_{max2} - t_{max1}$

TABLE 3

Model #1: Variation of the Capillary Length L1 between the Column End and the 1st Detector Cell. Flow Rate = 1.5 ml/min; r = 0.01143 cm; 6-Sigma-volume = 2.25 ml; Capillary #2: L2 = 50 cm.

Capillary Length L2		Apparent Dead Time Sec. (*)	Apparent -Geometric Dead Time Sec.	Asymmetry A			Analogue of Plate Number			Decrease of Max. Peak Height in 2. Cell(%)
cm	sec.			1.cell	2.cell	Increase %	1.cell P1	2.cell P2	Decrease %	
0	0	1.59	.77	1.000	.995	.52	1.000	.968	3.19	2.86
25	.41	1.26	.44	.998	.992	.55	.983	.955	2.83	2.61
50	.82	1.13	.31	.995	.990	.46	.986	.943	2.62	2.44
75	1.23	1.05	.23	.992	.988	.43	.955	.932	2.44	2.31
100	1.64	.99	.17	.990	.986	.42	.943	.921	2.32	2.22
150	2.46	.91	.09	.986	.982	.38	.921	.901	2.12	2.05
200	3.28	.85	.03	.982	.979	.38	.901	.884	1.98	1.92

(*) = $t_{max} - t_{max1}$

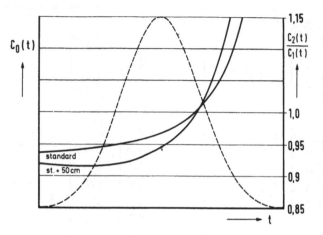

FIGURE 8
Variation of the Capillary Length Between the Detector Cells (L2 of
Model #1 = 50 cm and = 100 cm, Respectively, all other Values Equal
the Standard Conditions): Quotient Curves $C_2(t)/C_1(t)$ and the
Eluated Concentration Distribution $C_0(t)$. Correction of the Dead
Volume by the Exact Geometric Dead Volume.

is the relative decrease of the apparent dead volume between the
two detector cells with increasing capillary radius (Col. 5., Table
4).

Influence of the mixing vlume between the two detector cells

In a GPC-unit with multiple detection, two possibilities for a
dead volume with backmixing between the two detectors are present:
a "poor" capillary union "with high dead volume" and a mixing process occuring directly in a detector cuvette. In both the cases
the volume is about 0.01 ml. Inserting this value into model #2 results in surprisingly low additional effects (Table 5). We found
the same results for the quotient curves $c_2(t)/c_1(t)$. This observation can be explained by comparing the results for an isolated
mixing volume (model #2 with L1, L2, L3 = 0, V_{mix} = 0.010 ml) and
for an isolated capillary of the same volume (model #1 with L1=0 and
L2=24.38 cm, R=0.01143 cm). The results are shown in Table 5 and
Figure 9. The shape of the elution curve is more disturbed by the
capillary than by the mixing volume (e.g., the decrease of the ana-

TABLE 4

Model #1: Variation of the Radius R of the Capillaries. Flow Rate: 1.5 ml/min; 6-Sigma-volume: 2.25 ml; Capillaries: L1 = 25 cm, L2 = 50 cm.

Radius R cm	Capillaries L1 Sec.	L2 Sec.	Apparent Dead Time Sec.(*)	Apparent -Geometric Dead Time Sec.	Asymmetry A 1.cell	2.cell	Increase in 2.cell %	Analogue of Plate Number 1.cell P1	2.cell P2	Decrease %	Decrease of Max. Peak Height in 2.cell(%)
.01143	.41	.82	1.26	.44	.998	.992	.55	.983	.955	2.83	2.6
.02286	1.64	3.28	3.55	.27	.990	.975	1.51	.943	.867	8.02	7.8
.05	7.85	15.71	11.72	-3.99	.965	.924	4.18	.816	.643	21.22	20.8

TABLE 5

Model # 2: Results with Mixing Dead Volume in Comparison to Model #1. Flow Rate: 1.5 ml/min; Radius = .01143 cm; 6-Sigma-volume = 2.25 ml mixing volume: 0.010 ml.

Lengths of Capillaries L1 cm	L3 cm	L2 cm	Apparent Dead Time Sec.(*)	Apparent -Geometric Dead Time Sec.	Asymmetry A 1.cell	2.cell	Increase in 2.cell %	Analogue of Plate Number 1.cell P1	2.cell P2	Decrease %	Decrease of Max. Peak Height in 2.cell(%)
Model #2 with a mixing volume:											
25	25	25	1.998	1.177	.997	.992	+.53	.985	.952	3.34	2.88
Similar but without a mixing volume:											
25	0	50	1.26	0.44	.998	.992	+.55	.983	.955	2.83	2.61
"Dual-Detector":											
25	0	0	.3996	-.0004	.997	.998	-.02	.985	.984	.07	.04
Isolated Mixing Volume:											
0	0	0	.3996	-.0005	1.000	1.0002	-.018	1.000	.9993	.07	.035
Isolated Capillary (model #1):											
0	0	24.38	.904	.504	1.000	.998	+.20	1.000	.983	1.69	1.48

(*) = $t_{max2} - t_{max1}$

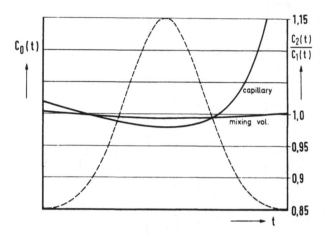

FIGURE 9

Effects of an Isolated Mixing Volume and of a Capillary of the Same
Size, Respectively. Geometric Dead Volume = 0.010 ml. Shifting of
$C_1(t)$ by the Calculated Apparent Dead Volume for the Evaluation of
the Quotient Curve $C_2(t)/C_1(t)$; --- = Input Concentration Curve
$C_1(t)$.

logue P of plate number and the size of apparent dead volume, Table
5). Thus, the differences between the experiments with a "good" and
a "poor" capillary union (Table 6, type B and D) should be interpreted by a model with a short capillary exhibiting a laminar flow profile and not by a model having a volume with perfect mixing.

CONCLUSION

Model calculations show that laminar flow with a parabolic flow
rate distribution is a reason for the experimentally observed deformation of the elution curves between the cuvettes in multiple
detector GPC-equipment. This model gives for example the same increase of apparent dead volume as the experiment. Further, the
lowest relative deformation in a given GPC-unit is found for elution curves with the lowest gradient of the detected property (c,
c.M or c.η).

An optimal value for the apparent dead volume can be determined by experiment only for a combination of two concentration detectors in the analysis of homopolymers. This is not possible for co-

TABLE 6

Apparent Dead Volume for a Series of Different Connections Between an UV-Detector and a Refractometer. GPC-Unit with 2 Columns 1 x 60 cm, Radius of Capillaries = .01143 cm; Solvent: THF; Flow Rate = 1.5 ml/min; Test Samples: Ethylbenzene and Polystyrene PCC 110000; Peak Maxima: 72.3 and 54.5 ml Respectively; Plate Number in 1st Detector (=UV): 35700 ± 300.

Connection Type	Plate Number in 2nd Detector	Apparent Dead Volume ml(*)	Apparent Dead Volume ml **)	Additional Geometric Dead Volume Relative to A ml	Additional Apparent Dead Volume Relative to A ml
A) Immediately	35000	.065	.060	–	–
B) A+ 50 cm Capillary	34800	.131	.083	.021	.066
C) A+ 100 cm Capillary	33700	.169	.120	.041	.104
D) A+ 50 cm Capillary and one normal union (=with high dead volume = .011 ml)	33500	.126	.095	.031	.061

*)   Measured by the peak difference of ethylbenzene

**) Measured by an optimation of the quotient curves for narrowly distributed polystyrenes

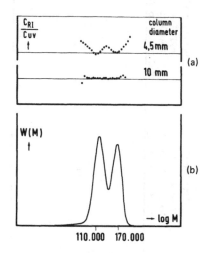

FIGURE 10

Multiple Detection of a Mixture of Two Narrowly Distributed Polystyrene Samples (PCC 110000 and PCC 173000): Quotient Curves $C_{RI}/$
C_{UV} Calculated with the Optimum Value for the Dead Volume Correction:
a) With 4 Column a 10 x 300 mm, Flow Rate = 1.5 ml/min;
   Peak Maxima: 51.9 and 54.7 ml
b) With 4 Columns a 4.5 x 300 mm, Flow Rate = .4 ml/min
   Peak Maxima: 7.8 and 8.2 ml.

polymer samples with composition being a function of the molecular
weight and is also not possible when molecular size detectors are
used.

Thus, we recommend the use of columns with a large diameter
for the multiple detection of narrowly distributed polymer samples.
The calculations show further a greater disturbance by a capillary
with a laminar flow characteristic than by a dead volume of the
same size with perfect mixing. The effect of a "poor" capillary
union (= with hig dead volume) is better described by a short capillary with parabolic flow than by a volume with perfect mixing.

REFERENCES

1. Runyon, J.F., et. al., J. Appl. Polym. Sci., _13_, 2359,(1969).

2. Cantow, H.J., et. al., Kaut. Gummi Kunstst., _21_/11, 609,(1968).
   Ref. in Chem. Abst., _70_/8, 29849 Q.

3. Grubisic-Gallot, Z., et. al., J. Appl. Polym. Sci., 16, 2931, (1972).

4. Ouano, A.C., J. Chromatogr., 118, 303,(1976).

5. Denbigh, K., Chemical Reactor Theory, Cambridge, 1965, p. 72.

6. Bronstein, I.N., and Semendjajew, K.A., Taschenbuch der Mathematik, Frankfurt am Main, 1972, p. 378.

AQUEOUS GEL PERMEATION CHROMATOGRAPHY: THE SEPARATION
OF NEUTRAL POLYMERS ON SILICAGEL

J.A.P.P. van Dijk, J.P.M. Roels and J.A.M. Smit

Gorlaeus Laboratories
The State University of Leiden
Leiden, The Netherlands

ABSTRACT

The application of aqueous GPC has been focused upon the system
of dextrans in water with silicagel as the column packing. Starting
from hetero-disperse standards calibration relationships have been
considered, which differ in the choice of the separation parameter.
The influence of dispersion on GPC calibration has been estimated.
Furthermore, the separation has been studied by measuring the number-
average molecular weight of dextrans obtained from a number of eluted
fractions. Finally, attention has been paid to the problematic solubility of silica, a corollary overlooked somewhat in literature*.

INTRODUCTION

Recently, we have studied the separation of dextrans by GPC,

using silicagel as the column packing and water as the solvent.

This chromatographic system appears to be a model for aqueous GPC

with neutral polymers[1] owing to its easy RI detectability and its

good analyzability with a variety of means.

In a former paper[2] we have shown, that even broad molecular-

---

*NOTE: After the submission of this paper we have taken note of the
work of Barker et al., that appeared very recently in the Journal of
Chromatography, 174, 143, 1979, and is dealing with problems associated with aqueous column packings. Their report agrees with some
of our findings.

weight distribution standards are useful to obtain an accurate relationship between the molecular weight M and elution volume v. The
only requirement is the availability of a number of samples, which
beforehand have been characterized by the weight-average molecular
weight $\overline{M}_w$ and the number-average molecular weight $\overline{M}_n$. In a subsequent paper[3] we have followed a similar route to arrive at a rela-
tionship between the intrinsic viscosity [η] and v. In this case
only the intrinsic viscosity of the heterodisperse samples [η] must
be known previously. In this paper we comprehend several ways of
calibrating using the separation parameters M, [η] and M [η], the
latter being the so-called universal parameter introduced by Benoit[4].

Apart from the rather indirect method cited above, calibration
can be performed directly by monitoring the eluate continuously for
the separation parameter of interest. On-line molecular weight and
viscosity detector systems have been introducted in GPC[5,6]. Al-
ternatively, one may practice the more laborous method of segmenting
the eluate followed by analyzing the segments afterwards[3,7]. However, these methods are not free of complications. First, the elution volume actually measured does not refer to a pure separation
process unless it has been corrected for dispersion. Another compli-
cation arises, when, as frequently met with aqueous GPC, the column
packing does not turn out to be inert. The packing material may adsorb the solute and/or may dissolve in the eluent. Solute adsorption can easily be recognized from the chromatograms by the change
of their shape and the shift of their elution range. Unfortunately,
the solubility of the column packing remains latent unless the eluate
is examined thoroughly. With respect to the latter, this paper reports that dissolved silica interferes with the determination of $\overline{M}_n$
of a number of collected GPC fractions of dextrans in water. Besides
the chromatograms of polyacrylamide in water are presented which
strongly indicate adsorption effects.

EXPERIMENTAL

Measurements involving column calibration and the determination
of zonal dispersion were performed on a Waters' Model 200 GPC, equip-

ped with 4 columns (diameter 3/8 inch) packed with deactivated silicagel (Porasil, code AX, BX, CX, and DX, Waters Assoc.) in the particle
size range of 75-125 µm. Chromatograms were obtained with the usual
RI detection and a siphon counting the elution volume in units of
0.73 ml. The operational conditions were as follows: solvent water;
solutes 8 dextran standards (Pharmacia) with specifications as listed
in Table 1; column temperature 30°C; flow rate, 1 ml/min; sample concentration 0.25%; injection time 120 s. Some measurements were carried out with PAAm (polyacrylamide) in water. In order to estimate
the dispersion correction the current technique of reverse-flow was
followed. The data thus obtained were processed on an IBM/360 computer according to a numerical method developed in our laboratory[10].
The dextran standards were characterized by measuring the intrinsic
viscosity [η] in an Ubbelohde viscometer at 30 $\pm$ 0.05°C.

Besides combined GPC-viscosity measurements were performed by
collecting and analyzing separate fractions. Since the elution
ranges of these fractions should meet exactly the corresponding
chromatogram sections, the elution volumes were corrected for the
small dead volume between the detector and the siphon-counter.

Additional fractionations on a preparative scale were required
in order to allow end-group analysis of the collected solutes. They
were carried out on a Waters' Model CD 100 Chromatoprep, in which 4
Porasil columns (diameter, 1 inch) were installed. The detection
system consisted of a Waters' R400 RI detector. The flow rate amounted 12 ml/min. Each run 20 ml of a 2% solution was injected. After
one recycling mode the solution was collected in 4 fractions. In
order to obtain the $\overline{M}_n$ data, the fractionated dextrans were subjected to the end-group analysis after Somogyi-Nelson[11,12] with iso-
maltose as the reference compound. Extinctions were measured on an
Eppendorf-spectrophotometer.

DISPERSION

GPC has grown to one of the most powerful techniques for the
determination of the molecular weight distribution of polymers.
However, it is somewhat disappointing that the actually measured

chromatogram does not reflect merely this distribution. This is
caused by the fact that the injected zone has a final width, which
enlarges during the passage through the columns. Consequently, each
chromatogram possesses a sort of elementary spreading as also can be
seen from the chromatogram of a monodisperse sample. This phenomenon,
known as zonal spreading or dispersion has engaged many authors. It
is first mathematically described by Tung[8] who has also pointed out
the reverse-flow technique by which dispersion can be estimated
separately[9]. Among the many attempts to numerically solve Tung's
dispersion equation, we have chosen our own method[10] for the cases
considered here. Typical examples are shown in Figures 1 and 2. Inspection of Figure 1 reveals, that the uncorrected chromatogram covers
a larger elution range than the corrected one does. As a consequence
one finds from the uncorrected elution cruves too low values for $\overline{M}_n$

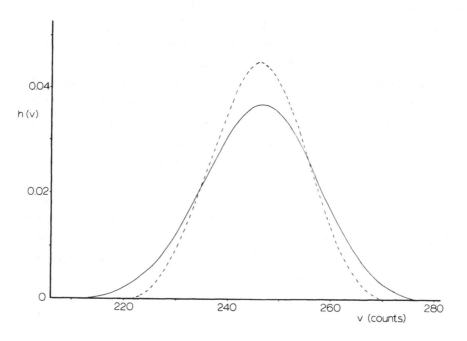

FIGURE 1

Normalized Chromatograms of Dextran T10. Solid Curve actually
Measured; Dotted Curve Corrected for Dispersion.

because in the domain of large elution volumes low molecular weight
material is counted, which is not present. Reversely, the corresponding value of $\overline{M}_w$ is too high, because in the domain of small elution
volumes erroneously high-molecular weight material is counted. This
dispersion effect on the values of $\overline{M}_n$ and $\overline{M}_w$ has been reported for
different systems[2,10]. Furthermore, it must be noted that disper-
sion is more pronounced in narrow peaks than in broad peaks and often
even neglible in the latter[10]. The integral form of the chromatogram (Fig. 2) is particularly suitable when a correction on the
elution volume is to be applied. For instance a certain weight fraction of polymer is eluted up to the elution volume v_e. The corrected
elution volume v_e' is then found as indicated in Figure 2. Obviously this correction is minimal around the top, but becomes more important near the edges of the chromatogram.

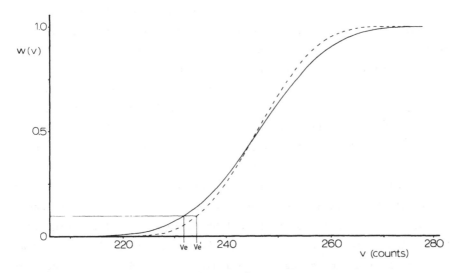

FIGURE 2
Normalized Chromatograms of Dextran T10 Transformed to Their Integral
Form. Solid Curve actually Measured; Dotted Curve Corrected for Dis-
persion.

CALIBRATION

The most convenient way of calibrating GPC is a standardization
with a number of monodisperse samples. Though the height of their
narrow peaks depends strongly on dispersion, their position does not
and the peak elution volumes may be related directly to the separation parameter concerned. However, heterodisperse standards yielding broad peaks, may also be used for calibration, which is shown
here using the system dextran, water and silicagel. Dextrans origi-
nating from the same bacterial strain form a homologous series de-
spite of their branching[3]. As such they behave as linear polymers.

Let us regard the calibration procedure to be followed in view
of eight dextran standards previously characterized by $\overline{M}_w$ and $\overline{M}_n$.
Our aim is to find the elution volumes conjugate to each of the form-
er sixteen values of the molecular weight. This can be done by a
method of iteration, which starts with an approximately chosen value
$\overline{v}_e$ for each dextran standard and ends in a stationary final value
v_e^*. We define $\overline{v}_e$ as the weighted-average value of the elution
volumes of the chromatogram, in which the normalized heights have
been taken as the weights. As the first step a calibration relationship is established between the eight preset values of $\overline{M}_w$ and the
starting values $\overline{v}_e$ by a four term polynomial approximation. The
crude calibration relation just obtained enable us to calculate new
values of $\overline{M}_n$ and $\overline{M}_w$ from the chromatograms and to gather the conjugate elution volumes. The latter are combined again with the preset
values of $\overline{M}_w$ and of $\overline{M}_n$, yielding a second and better calibration
polynomial. Continuing the iteration we observe that the succeeding
values of the elution volume converge rapidly (Fig. 3). Simultaneously the four coefficients of the calibration polynomial converge to
their final value. Once arrived at the final calibration polynomial,
one can recalculate for each chromatogram $\overline{M}_w$ and $\overline{M}_n$. These recalculated values ought to agree with the preset values.

The foregoing iteration method is not restricted to molecular
weight averages, but applies similarly to viscosity averages. Figure
3 shows that even a better convergence has been attained in the

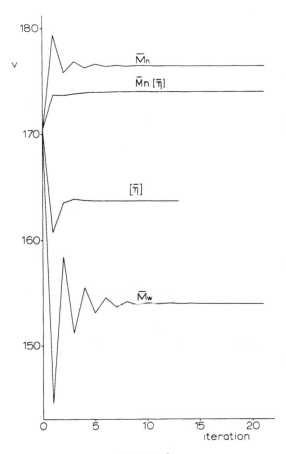

FIGURE 3
Illustration of the Convergence of the Iteration Method in View of
the Elution Volume.

iteration where the intrinsic viscosity is involved. Clearly, here
$\bar{v}_e$ (170 counts) is a favorably chosen starting value, as it is already close to the final value v_e^* (164 counts).

In principle, the iteration method is also useful for a universal calibration based on the parameter $M[\eta]$, albeit in a slight different form[2]. In this case the standards have to be characterized
previously by the product $\bar{M}_n[\bar{\eta}]$, which is equal to the number-average
value of the universal parameter[2]. Also here convergence has been

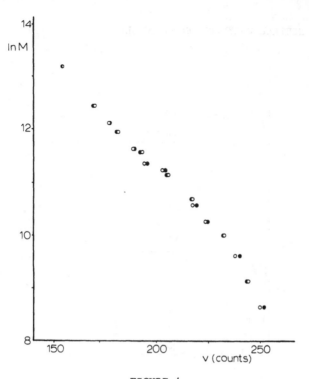

FIGURE 4

Calibration Relationship Between M and v Corrected (O) and Uncorrected (●) for Dispersion.

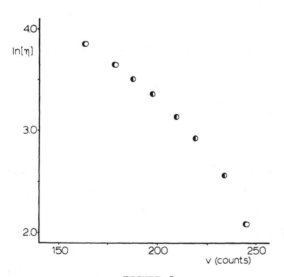

FIGURE 5

Calibration Relationship Between [η] and v Corrected (O) and Uncorrected (●) for Dispersion.

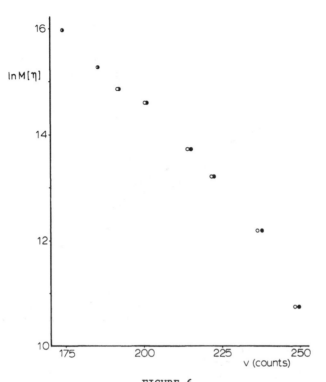

FIGURE 6
Calibration Relationship Between M[η] and v Corrected (O) and Uncorrected (●) for Dispersion.

found (Fig. 3). It can be shown, that other starting conditions[2]
lead to identical results.

The procedure outlined above has led to the establishment of
three calibration relationships, shown as discrete points in Figures
4-6. The full dots refer to points obtained from uncorrected chromatograms, the open circles refer to corrected chromatograms. For the
sake of clarity the calibration curves have not been drawn in the
figures. Apparently dispersion influences marginally the calibration
relationships, though an increasing tendency is observed for the respective parameters [η], M[η] and M. This becomes clear by considering the shift of the elution volumes due to dispersion. The elution
volume conjugated to [η] is found in an elution range around the

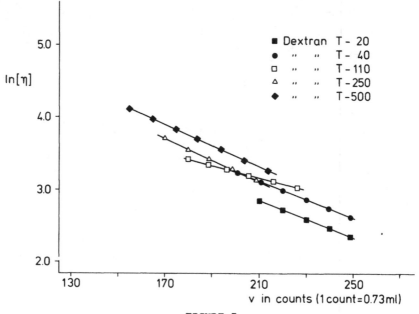

FIGURE 7

ln[η] Versus v (Uncorrected)
■, Dextran T 20; ●, Dextran T 40;
□, Dextran T110; △ , Dextran T250;
◆, Dextran T500.
(Reproduced from the J. Polym. Sci., Courtesy of John Wiley and Sons.)

peak, where the dispersion effect is minimal. However, the elution
volumes conjugated to $\overline{M}_w$ and $\overline{M}_n$ are found more near the limits of
the chromatograms where the dispersion effect plays a major part.

Recalculated values of the relevant parameter are collected in
Table 1. Agreement is present between the calculated and the preset
values, although the dispersion correction does not always imply an
improvement. Again it is seen, that dispersion hardly affects the
data involving [η] and M[η].

Contrary to the cases regarded above dispersion emerges more as
a disturbing effect when separate fractions are collected during a
GPC run. This situation is pictured in Figure 7. For 5 dextran

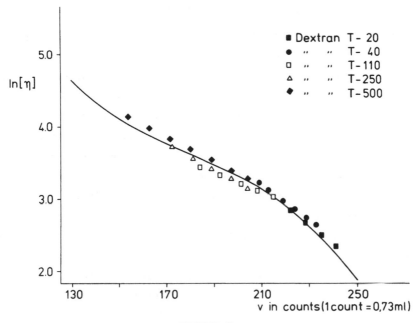

FIGURE 8

ln[η] Versus v (Corrected)
■, Dextran T 20; ●, Dextran T 40;
□, Dextran T110; △, Dextran T250;
◇, Dextran T500.
(Reproduced from the J. Polym. Sci., Courtesy of John Wiley and Sons.)

standards, linear relationships were found between the logarithm of
the measured intrinsic viscosity of the fractions and their actually
measured elution volumes. However, the points in Figure 7 do not fit
a single curve. Let us see how the discrepancies in Figure 7 vanish
by application of the dispersion correction. The latter gives rise
to a contraction of the elution range (cf. Fig. 1 and 2). At the
limits of the chromatogram the elution volumes become respectively
larger and smaller, whereas they remain unchanged in the middle part.
As result the lines in Figure 7 are rotated clockwise. Figure 8
shows the ameliorated situation, in which the points fit a single
calibration curve. This solid curve has been prepared by a least-
squares fit through the open circles of Figure 5.

ACTIVITY OF THE COLUMN PACKING

In spite of many favorable aspects with respect to separation and detection, aqueous GPC is lacking the provision of completely inert column-packing material. For silicagel two sources give rise to interference with the separation process, i.e, its solubility and its active surface enhancing adsorption. The solubility of silicagel remains usually unnoticed in the analytical GPC, where the differential detection technique eliminates the effect of contamination to a certain degree. As long as the dissolved silicagel does not interact with the polymer to be separated no serious problem is met here. However, dealing with preparative GPC in which the fractionated material is isolated, one is left with a product contaminated with column material.

We have investigated the effect by analyzing water which has been passed through columns filled with Porasil and µBondagel supplied by Waters. Freeze-drying of both specimens of water yielded residues whose IR-spectra coincide completely with the reference IR-spectrum of solid Porasil BX. Another experiment was more remarkable. For dextran T70, $\overline{M}_n$ was determined from the end-group analysis in two circumstances: before injection and after elution. The value of $\overline{M}_n$ of the "treated" sample amounted to 30400 which is much lower the the value 40000 obtained for the "untreated" sample. Correspondingly $\overline{M}_n$ for "untreated" dextrand T40 came out on 29480. Thus, the values found for incomtaminated dextrans are in excellent agreement with the data given by Pharmacia (cf. Table 1). However, apparently the value of $\overline{M}_n$ has been pushed down by interfering dissolved silicagel. The weak point here is the choice of the reference solution for the analysis. It seems better to take the mobile phase of GPC as reference instead of pure water, though even then, there is no guarantee that one exactly attaims the Si content of the isolated dextran fractions. If the latter is increased in the reference solution, indeed higher values of $\overline{M}_n$ can be obtained, which even approach the expected values (Table 1), however, at improbably high Si concentrations. Evidently

TABLE 1

GPC Standardization Data of Dextrans in Water.

Sample	Pharmacia $\bar{M}_w \times 10^{-5}$	GPC-uncorrected $\bar{M}_w \times 10^{-5}$	$\bar{v}_e$ (Counts)	v_e^* (Counts)	GPC-corrected $\bar{M}_w \times 10^{-5}$	$\bar{v}_e$ (Counts)	v_e^* (Counts)
T500	5.32	5.38	170.4	153.7	5.46	170.5	154.0
T250	2.53	2.48	182.6	168.7	2.40	182.8	169.6
T150	1.54	1.56	189.6	180.3	1.51	189.5	181.2
T110	1.06	1.06	198.2	191.7	1.01	198.2	192.9
T 70	0.695	0.689	209.7	204.5	0.661	209.7	205.5
T 40	0.444	0.442	219.0	216.6	0.421	219.1	217.4
T 20	0.223	0.213	234.3	232.1	0.200	234.0	232.6
T 10	0.093	0.106	246.1	243.4	0.093	245.9	244.1

Sample	Pharmacia $\bar{M}_n \times 10^{-5}$	GPC-uncorrected $\bar{M}_n \times 10^{-5}$	$\bar{v}_e$ (Counts)	v_e^* (Counts)	GPC-corrected $M_n \times 10^{-5}$	$\bar{v}_e$ (Counts)	v_e^* (Counts)
T500	1.83	1.76	170.4	177.1	1.80	170.5	176.5
T250	1.125	1.16	182.6	189.0	1.18	182.8	188.3
T150	0.860	0.933	189.6	195.4	0.975	189.5	193.9
T110	0.760	0.695	198.2	204.3	0.730	198.2	202.6
T 70	0.395	0.399	209.7	219.0	0.424	209.7	217.2
T 40	0.289	0.309	219.0	224.8	0.324	219.1	223.3
T 20	0.150	0.134	234.3	240.0	0.146	234.0	237.7
T 10	0.057	0.056	246.1	251.8	0.059	245.9	249.7

TABLE 1 (Continued)

Sample	$[\bar{\eta}]_{exp}$ (ml/g)	GPC-uncorrected			GPC-corrected		
		$[\bar{\eta}]$ (ml/g)	$\bar{v}_e$ (Counts)	v_e^* (Counts)	$[\bar{\eta}]$ (ml/g)	$\bar{v}_e$ (Counts)	v_e^* (Counts)
T500	47.20	47.49	170.4	163.1	47.50	170.5	163.8
T250	38.24	37.76	182.6	178.3	37.67	182.8	179.0
T150	33.18	33/16	189.6	187.5	33.24	189.5	187.7
T110	28.76	28.64	198.2	197.4	28.65	198.2	197.5
T 70	22.91	23.24	209.7	209.4	23.25	209.7	209.3
T 40	18.63	18.99	219.0	218.8	18.92	219.1	218.9
T 20	12.98	12.48	234.3	233.6	12.54	234.0	233.6
T 10	8.01	8.15	246.1	244.9	8.13	245.9	245.2

Sample	Pharmacia/exp $\bar{M}_n[\bar{\eta}] \times 10^{-6}$ (ml/g)	GPC-uncorrected			GPC-corrected		
		$M_n[\eta] \times 10^{-6}$ (ml/g)	$\bar{v}_e$ (Counts)	v_e^* (Counts)	$M_n[\eta] \times 10^{-6}$ (ml/g)	$\bar{v}_e$ (Counts)	v_e^* (Counts)
T500	8.638	8.620	170.4	174.0	8.630	170.5	174.1
T250	4.302	4.316	182.6	185.4	4.262	182.8	185.5
T150	2.853	3.020	189.6	192.2	3.058	189.5	191.6
T110	2.186	1.935	198.2	201.2	1.947	198.2	200.5
T 70	0.905	0.932	209.7	215.1	0.935	209.7	214.1
T 40	0.538	0.597	219.0	222.4	0.583	219.1	221.6
T 20	0.195	0.175	234.3	237.8	0.179	234.0	236.3
T 10	0.046	0.047	246.1	249.6	0.047	245.9	248.3

the choice of the reference solution must be made carefully in all cases involving on-line detection of separation parameters.

As we have discussed already, collected fractions are adequately characterized on the basis of the product $\overline{M}_n[\overline{\eta}]$. From this point of view it seems worthful to investigate the possibilities of measuring $\overline{M}_n$ of dextrans fractioned by GPC. Hitherto, we have only succeeded in performing experiments on a preparative scale. Their results are listed in Table 2. For T70 the data have been gained from 7 experiments and presented as average values. The data of T40 originate from a single experiment. A correction for the interference of dissolved silicagel has been applied, the dispersion correction has been omitted. Table 2 shows that separation of dextrans in GPC can be observed by measuring $\overline{M}_n$ of the eluted fractions. Dextran T70 has an average elution volume of 2009 ml which indicated a value of $\overline{M}_n$ in the range 21000 to 45000 (see Table 2 cf. Table 1). Analogously one find for T40 with $\overline{v}_e$ equal to 2048 ml, a value of $\overline{M}_n$ between 23000 and 32500 (see Table 2 cf. Table 1).

Finally, we discuss briefly the phenomenon of adsorption of polyacrylamide on silicagel (deactiviated Porasil). Our findings seem to be in contrast with other reports[1,13] which suggest that

TABLE 2

Fractionation of Dextrans on GPC. The Separation Parameter is $\overline{M}_n$. $\overline{v}_e$ is the Average Elution Volume of the Collected Fraction.

Fraction Number	T-70		T-40	
	$\overline{M}_n$	$\overline{v}_e$ (ml)	$\overline{M}_n$	$\overline{v}_e$ (ml)
1	121400	1714	75800	1640
2	45000	1904	32500	1897
3	21000	2092	23000	2097
4	9600	2304	13900	2326

polyacrylamide can be chromatographed successfully on silicagel with water as the eluent. Figure 9 shows a sequence of chromatograms, obtained by subsequent runs of the same sample. The following remarks may be made:

1) In the course of time the chromatograms shift to the domain of smaller elution volumes (Fig. 9).

2) There is a remarkable change of shape of the elution curves (Fig. 9).

3) Especially in the first chromatograms of the sequence, the surface under the curves increases.

4) Freshly prepared samples do not disturb the picture, their chromatogram fits completely the sequence.

5) The chromatogram of dextrans neither change nor shift during the sequence of polyacrylamide-runs.

Hence, one may conclude that in some way the column packing has gotten modified properties by the adsorption of polyacrylamide, which do not affect the separation of dextrans.

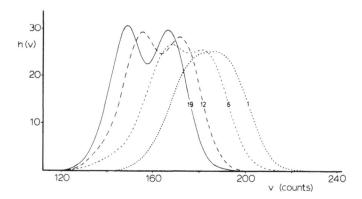

FIGURE 9
Illustration of Adsorption Effects of Silicagel on Polyacrylamide in Water during a Sequence of GPC Runs.

REFERENCES

1. Cooper, A.R., and Van Derveer, D.S., J. Liquid Chromatog., 1, 693 (1978).

2. Vrijbergen, R.R., Soeteman, A.A., and Smit, J.A.M., J. Appl. Polym. Sci., 22, 1267 (1978).

3. Soeteman, A.A., Roels, J.P.M., van Dijk, J.A.P.P., and Smit, J.A.M., J. Polym. Sci., Polym. Phys. Ed., 16, 2147 (1978).

4. Benoit, H., Grubisic, Z., Rempp, P., Dekker, D., and Zilliox, J.G., J. Chim. Phys., 63, 1507 (1966).

5. Ouano, A.C., and Kaye, W., J. Polym. Sci., Part A-1, 12, 1151 (1974).

6. Ouano, A.C., J. Polym. Sci., Part A-1, 10, 2169 (1972).

7. van Dijk, J.A.P.P., Henkens, W.C.M., and Smit, J.A.M., J. Polym. Sci., Polym. Phys. Ed., 14, 1485 (1976).

8. Tung, L.H., J. Appl. Polym. Sci., 10, 375 (1966).

9. Tung, L.H., Moore, J.C., and Knight, G.W., J. Appl. Polym. Sci., 10, 1261 (1966).

10. Smit, J.A.M., Hoogervorst, C.J.P., and Staverman, A.J., J. Appl. Polym. Sci., 15, 147 (1971).

11. Somogyi, M., J. Biol. Chem., 160, 61 (1945).

12. Nelson, N., J. Biol. Chem., 153, 375 (1944).

13. Hamielec, A.E., and Abdel-Alim, A.H., J. Appl. Polym. Sci., 18, 297 (1974).

UTILIZATION OF MULTIDETECTOR SYSTEM
FOR GEL PERMEATION CHROMATOGRAPHY

Z. Gallot

Centre de Recherches sur les Macromolecules
CNRS
Strasbourg, France

INTRODUCTION

It is well known that Gel Permeation Chromatography is one of the best tools for fast characterization of polymers. With the aim of increasing the possibilities of the usage of GPC, we have coupled this technique with a multidetector system[1-3] comprising an automatic viscometer, a UV spectrophotometer, an automatic densimeter and a light scattering photometer. In the present paper, we will describe the results we have obtained and which concern linear or branched homopolymers as well as copolymers.

EXPERIMENTAL

The classical Waters Assoc. GPC-200 apparatus having a 5 mL syphon and a 5 column set (Styragel) has been used. In all experiments, tetrahydrofuran or dimethylformamide were used as the solvent, the flow rate of which was adjusted to 1 mL/min at room temperature. The outlet of the syphon was connected to an automatic viscometer having a capillary of 0.5 mm in diameter and a length of 20 cm. The diameter and the length of the capillary were selected so as to have a flow time shorter than the time required for the consequent filling of the syphon. The temperature of the viscometer was controlled to ± 0.01°C. The flow times were determined by an automatic device which has been commercialized by Fica.

The UV spectrophotometer (Varian 635) and the automatic digital densimeter (DMAO 2) conceived by Dratky[4] have been placed in a series after the refractometer. Let us recall that the principle of this densimeter is based on the measurement of the vibration of a glass U-tube filled with a liquid of unknown density, the vibration period T being proportional to the polymer concentration.

The light scattering detector was conceived in our laboratory from a Fica 42000 apparatus, using a helium-neon laser (632,8 nm) as a light source and a flow cell which allows the continuous measurement of the scattered intensity. Depending on the experiments, the volume of the cell was between 0.3 mL and 3.5 mL. The cell was put after the refractometer, so that the dead volume between the two detectors is the lowest.

RESULTS AND DISCUSSION

In a first step, we have employed the coupling GPC-automatic viscometer to determine, by using the universal calibration, the polydispersity of linear or branched homopolymers. The results concerning the linear polystyrenes and linear or branched polyvinylacetates are listed in Table 1. It can be seen that the values of intrinsic viscosity and molecular weight obtained by the coupling GPC-multidetector system are in good agreement with the corresponding values determined by the classical methods of characterization.

We have also characterized the degree of branching which in the case of branched polymers can be defined as: $g' = [\eta]_{br} \big/ [\eta]_{1}$ where $[\eta]_{br}$ is the intrinsic viscosity of the branched polymer and $[\eta]_{1}$ is the intrinsic viscosity of the linear polymer of the same molecular weight. Table 2 illustrates the results obtained for star polystyrenes. The average branching indices determined by GPC and by classical techniques are quite similar.

We have been able to evaluate as well the distribution of g' as a function of molecular weight as it is shown in Figure 1 for a sample of polybutadiene (Cariflex BR-1220).

TABLE 1

Comparison of Intrinsic Viscosities and Weight Average Molecular
Weights Obtained by Classical Methods (Ubbelhode Viscometer, Light
Scattering) and with the Proposed Technique in Case of Homopolymers.

Sample	$[\eta]_{Ubbelohde}$	$\overline{M}_{w_{LS}}$	$[\eta]_{GPC}$	$\overline{M}_{w_{GPC}}$
PS (linear)	11.5	14,500	11.8	15,400
PS (linear)	66.4	173,000	64.3	182,000
PVAC (linear)	36.9	66,500	37.9	73,700
PVAC (linear	51.6	109,000	53.7	109,800
PVAC (linear)	63.6	143,000	61.2	158,000
PVAC (linear)	80.4	192,000	81.7	195,000
PVAC (branched)	31.2	48,000	26.1	51,700
PVAC (branched)	33.1	60,400	32.0	59,400
PVAC (branched)	89.5	318,000	85.5	269,000
PVAC (branched)	112.0	500,000	111.7	501.300

TABLE 2

Determination of the Branching Index of Polystyrene of Well-Known
Structure.

Sample	Shape	$M_w \cdot 10^{-3}$ GPC	$M_n \cdot 10^{-3}$ GPC	M_w/M_n	g'_{GPC}	g'_{visco}
PS-1	star	90.9	84.8	1.07	0.69	0.73
PS-2	star	100.5	96.0	1.04	0.29	0.30
PS-3	star	272.0	255.0	1.07	0.54	0.50
PS-4	star	434.0	358.2	1.06	0.45	0.44
PS-5	star	538.0	503.5	1.07	0.31	0.36
PS-6	comb	134.0	121.0	1.10	0.34	0.33
PS-7	comb	254.0	226.0	1.12	0.31	0.30
PS-8	comb	346.0	323.0	1.07	0.25	0.30
PS-9	comb	423.0	324.0	1.30	0.39	0.47
PS-10	comb	846.0	561.0	1.51	0.45	0.47

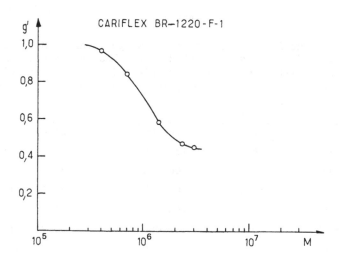

FIGURE 1

g' as a Function of Molecular Weight for a Sample of Polybutadiene.

In the case of copolymers which exhibit polydispersity in
molecular weight and composition, the fractions collected from the
chromatograph have different compositions, and thus the response
obtained from the refractometer is not only proportional to the
instantaneous concentration since it depends also on the composition. Under these conditions, determination of compositional polydispersity requires the utilization of an additional detector such
as a UV or IR spectrophotometer. In Table 3, the values of the
average molecular weights and intrinsic viscosities obtained with the
help of this device coupled with an UV spectrophotometer are given.
The good agreement between the various results indicated that the
present device is suitable for such studies. However, it should be
mentioned that the values of the limiting viscosity number differ beyond the experimental error for PS-PIP-3. The difference can perhaps
be attributed to the fact that the sample has a very broad distribution in compositional heterogeneity as well as in molecular weight.

It should be noted that, even for a homopolymer, it is not always possible to use as a detector, a differential refractometer or
a spectrophotometer. For example, when the value of refractive in-

dex increment is very small and when the polymer has no adsorption
band in a region where the solvent is transparent. In this particular case, we employed as a detector a Kratky densimeter. In order to show that in such a case this detector can be very useful, we
have studied a sample of polydimethylsiloxane in tetrahydrofuran.
We have obtained a well defined peak (Fig. 2) with our densimetric
detector and no signal at all by refractometry since in this solvent the dn/dc of PDMS is practially zero. On the other hand, we
have compared the results relative to a series of polystyrene samples obtained with a densimeter and with a classical refractometer.
The results (Table 4) are in good agreement.

All the results reported concern systems for which the solubility parameters of polymers, solvent and gel are similar. For such
systems the steric exclusion is the only separation mechanism and
utilization of the universal calibration is valid[5]. In the case
of systems including polymers, solvents and gels of different
polarities the universal calibration cannot be used[6]. Comparison
of the results concerning highly polar polymers such as poly-2-
vinylpyridine and polystyrene eluted in DMF on PS-gel shows that
two different curves are obtained (Fig. 3). The behavior of PS and
P2VP samples can be explained as follows: in the system PS/DMF/
Styragel, there are significant interactions between PS and Styragel
resulting in the slower elution of the polymer. In the second system, P2VP/DMF/Styragel, strong interaction between DMF and polar
P2VP can be assumed, so P2VP is "repulsed" from the gel. P2VP molecules are therefore eluted earlier than expected from the steric exclusion mechanism only, and such an effect is manifest in the steep-
er slope of the calibration curve. It must be emphasized that the
utilization of the detectors mentioned above requires the use of a
calibration curve. For this reason, we have couple GPC with a light
scattering detector, which provides the values of the weight-average
molecular weight without the help of a calibration curve.

TABLE 3

Comparison of Viscosities and Molecular Weights Obtained by Classical Methods and by GPC-Viscosities Measurements in the Case of Copolymers.

Sample	% PS UV	$\overline{M}_w$ LS	$\overline{M}_n$ OSM	$\overline{M}_w$ GPC	$\overline{M}_n$ GPC	$[\eta]_{Ubbelohde}$	$[\eta]_{GPC}$
PS-PIP-1	11.0	101,000	95,800	104,000	84,000	82	75
PIP-PS-2	52.7	159,000	137,700	167,500	136,200	103	100
PS-PIP-3	53.8	1,074,000	–	1,147,000	375,200	414	371
PIP-PS-4	93.4	114,000	88,100	124,500	89,000	57	54
PS-PMM-5	49.7	404,000	345,000	442,000	268,900	101	104
PS-PIP-PS-6	75.3	119,000	107,600	132,000	105,000	64.0	60.7
PS-PIP-PS-7	77.4	69,000	61,100	74,000	59,000	42.0	39.7

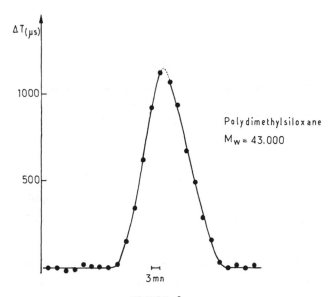

FIGURE 2

Densitometric Detection of Polydimethylsiloxane.

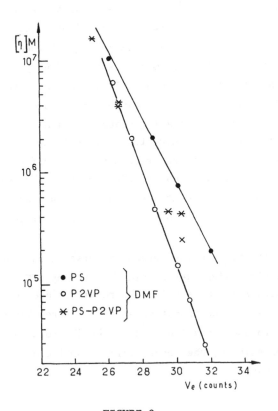

FIGURE 3

Elution Behavior of Polar Polymers and Polystyrene in DMF on a Poly-styrene Gel.

TABLE 4

Comparison of Data Obtained by Refractometer and Densimeter.

	Refractometer			Densimeter		
Sample	$\bar{M}_w$	$\bar{M}_n$	$\bar{M}_w/\bar{M}_n$	$\bar{M}_w$	$\bar{M}_n$	$\bar{M}_w/\bar{M}_n$
1	10,700	9,950	1.08	10,700	9,910	1.08
2	52,900	48,100	1.10	61,600	57,000	1.08
3	101,700	94,000	1.10	118,500	106,500	1.11
4	426,000	375,000	1.13	499,000	407,000	1.23

We have studied a series of narrow polystyrenes whose molecular weights were between 3,000 and 150,000. For these samples we have observed no disymetry in the angular dependence of the scattered intensity and, therefore, we have measured this parameter only for a scattering angle of $90°$. Figure 4 shows the responses given, respectively, by the refractometer and the light scattering photometer in the case of a polystyrene sample ($M_w = 51,000$). Table 5 summarizes the results obtained with a series of polystyrene samples by using ethylacetate as the eluent. It can be noted that the results provided by classical light scattering measurements and by GPC coupled with a light scattering photometer are similar. These results are not surprising since ethylacetate is not a good solvent for polystyrene. Therefore, the second virial coefficinet is low and the C/I values are practically independent of the polymer concentration. Under these conditions it is not necessary to extrapolate to zero concentration to obtain the weight-average molecular weight.

On the contrary, when the tetrahydrofuran is used as the solvent, the molecular weight determined by coupling GPC-LS depends on polymer concentration (Table 6). This result is due to the fact that the molecular weight is not obtained by extrapolation of the C/I ratio to zero concentration, and as tetrahydrofuran is a good solvent for polystyrene, the second virial coefficient must be taken into account.

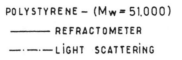

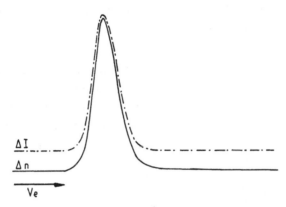

FIGURE 4

Polystyrene: Refractometry vs. Light Scattering.

TABLE 5

Comparison of Molecular Weight Obtained by Classical Light Scattering and GPC Coupled with a Light Scattering Photometer in Ethylacetate.

Sample	$\overline{M}_w$ LS	$\overline{M}_w$ GPC-LS
Polystyrene	4,000	3,900
Polystyrene	10,500	10,700
Polystyrene	25,400	24,700
Polystyrene	61,000	61,200
Polystyrene	130,000	134,000

TABLE 6

Influence of Polymer Concentration on Molecular Weight for a Polystyrene Sample ($\overline{M}_w$ = 125,000) in Tetrahydrofuran.

$C(g/mL) . 10^4$	$\overline{M}_w$
4.3	107,000
7.2	96,500
9.9	90,500

Figure 5 illustrates the different responses given by the following detectors: refractometer, UV spectrophotometer, light scattering photometer and viscometer coupled with a GPC chromatograph for a polystyrene-polyisoprene block copolymer $(\overline{M}_w = 114,000)$ using tetrahydrofuran as the eluent.

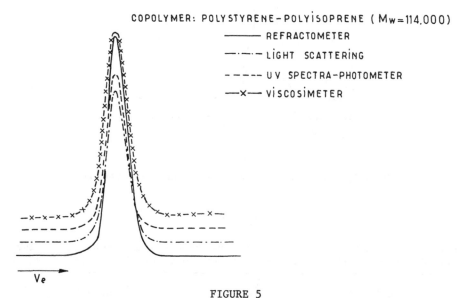

FIGURE 5

Polystyrene/Polyisoprene Copolymer: Various Detectors.

REFERENCES

1. Grubisic-Gallot, Z., Picot, M., Gramain, Ph., and Benoit, H., J. Appl. Polym. Sci., 16, 2931 (1963).

2. Francois, J., Jacob, M., Grubisic-Gallot, Z., and Benoit, H., J. Appl. Polym. Sci., 22, 1159 (1978).

3. Strazielle, C., and Gallot, Z., paper in preparation.

4. Kratky, O., Leopold, H., and Stalinger, H., Angew. Phys, 27, 273 (1969).

5. Dawkins, J.V., Maddock, J.W., and Coupe, D., J. Polym. Sci., A-2, 8, 1803 (1970).

6. Mencer, H.J., and Grubisic-Gallot, Z., J. Liquid Chromatogr., 2, 5, 650 (1979).

MOLECULAR CHARACTERIZATION OF DEGRADED POLYMERS

K. B. Abbås

Materials Laboratory
Telefonaktiebolaget L M Ericsson
Stockholm, Sweden

ABSTRACT

This paper discusses how molecular data, obtained by gel permeation chromatography (GPC), may be utilized for the elucidation of degradation kinetics. All procedures applied required accurate determinations of molecular weight averages or molecular weight distributions. Two systems were investigated in detail: (1) the repetitive extrusion of polycarbonate in a capillary rheometer and (2) the freezing and thawing of solutions of ultrahigh molecular weight polystyrene. Molecular scission was evaluated according to a method devised by Scott. It compares the relative decreases in M_w and M_n with those expected from theory. The nature of chain fracture was further elucidated by applying a general equation for scission kinetics proposed by Simha. This analysis made it possible to determine the scission frequency as a function of the position in the macromolecule.

INTRODUCTION

Many polymer properties are strongly dependent on molecular weight. As a consequence, chain scission usually causes detrimental effects on physical properties. This has been shown in many studies, where polymers have been exposed to various sources of energy. The photodegradation of polycarbonate is here given as an example[1]. A polycarbonate film with an $\overline{M}_w$ of 40,6000 and an $\overline{M}_n$ of 9,140 was degraded in a Xenotester for up to about 3,000 hr. Brittle fracture was observed for $\overline{M}_w$ <25,000 and $\overline{M}_n$ <6,000 which corresponded to ex-

posure times exceeding 700 hr. Changes in molecular structure do
not only occur during aging, but may also be observed during processing as well. At high processing temperatures, we see thermal as
well as thermooxidative degradation.

It is obvious that there is a great interest in finding reliable methods for the characterization of related reactions. The introduction of gel permeation chromatography (GPC) about 15 years ago
was a very important step in this development, although its potential
has not yet been fully exploited. It is the purpose of this paper to
discuss a number of procedures for the elucidation of degradation
kinetics. All methods assume accurate and reproducible molecular
weight determinations.

When poly-(vinyl chloride) is exposed to various forms of energy, hydrogen chloride is evolved. This reaction is accompanied by
discoloration which is explained by the development of conjugated
double bonds. The thermal degradation of PVC in nitrogen at 190°C
is here chosen to demonstrate the effect of branching reactions on
the molecular weight changes[2]. Molecular weights and molecular
weight distributions were calculated by a computer program devised
by Drott and Mendelson[3]. From a straight forward GPC analysis, we
obtained data as indicated by the circles in Figure 1. $\bar{M}_w$ increased
from 112,000 to 163,000 after 1.0% dehydrochlorination. No insoluble
material was obtained at this stage, but at 1.5% conversion about
6% insoluble material remained after treatment in THF at 120°C for
3 hr. This indicated that branching reactions occurred. Consequently, molecular weights were not correctly determined, as the calibration assumed linear polymer chains.

In combination with intrinsic viscosity data, GPC may be utilized for the determination of long chain branching. The ratio of
the intrinsic viscosity of a branched molecule to that of a linear
molecule having the same molecular weight has been related to the
branch points through the function g as follows:

$$[\eta]_{br} / [\eta]_1 = g^b \tag{1}$$

where b depends on the branching model that is used.

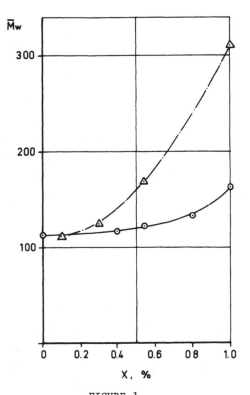

FIGURE 1

Weight Average Molecular Weight, $\bar{M}_w$, as a Function of Conversion, x;
(o) Uncorrected, (Δ) Corrected for Long Chain Branching. Data for
Poly(vinyl chloride) Degraded in Nitrogen at 190°C.

According to Zimm and Kilb[4], a value of b = 0.5 is suggested.
For polydisperse samples, the g-value is related to the weight-
number of trifunctional branch points in the following way[4]:

$$\left\langle g_3 \right\rangle_w = \frac{6}{n_W} \left[0.5 \left(\frac{2 + n_W}{n_W} \right)^{0.5} \ln \frac{(2 + n_W)^{0.5} + n_W^{0.5}}{(2 + n_W)^{0.5} - n_W^{0.5}} - 1 \right] \qquad (2)$$

The corresponding relationship for tetrafunctional branches is
expressed as:

$$\left\langle g_4 \right\rangle_w = 1/n_W \cdot \ln (1 + n_W) \qquad (3)$$

The number of branches were almost constant up to 0.3% dehydrochlorination[2]. At 0.54% it was tripled and at 1.0% it was about 11 times higher than the original value.

A comparison of $[\eta]_1$ and $[\eta]_{br}$ makes it possible to determine molecular weights which are corrected for long chain branching. The results are shown in Figure 1 (triangles). It clearly demonstrates the effect of branching on the molecular weight determinations - independent of the fact that data were obtained from two separate experiments[2,5]. At 1.0% dehydrochlorination the "true" $\overline{M}_w$ was almost twice as high as the uncorrected value. This also meant that the onset of molecular weight increase was observed at an earlier stage. These calculations were carried out assuming trifunctional long chain branching.

CRITERIA FOR RANDOM SCISSION

Scission reactions in polymers might occur at specific sites (e.g., at weak links) or randomly, but other mechanisms are also possible. Consequently, Bueche predicted that degradation induced by higher shear stresses causes rupture near the middle of chains, where the extending forces are at maximum[6]. To the best of our knowledge experimental results, showing fracture of macromolecules exactly in the middle, have not been published. However, data have been reported, which complies with a theory predicting a Guassian distribution for the location of rupture centered around the middle of the polymer chain[7,8]. The MWD of a polymer undergoing random scission approaches the most probable distribution, which is characterized by a heterogeneity index ($H = \overline{M}_w/\overline{M}_n$) of two. A plot of H as a function of the extent of degration has thus often been employed as a test for random scission. This requires accurate values of molecular weights as well as high extents of degradation.

As an example of such data, Table 1 shows the molecular weight changes as a result of freezing and thawing of ultrahigh molecular weight polystyrene in p-xylene[9]. At 2.5 g/1, the heterogeneity index increased from 1.1 to 1.8 after 45 freezing cycles. This indicates a random scission reaction, but although the extent of

TABLE 1

Molecular Weight Changes as a Result of Freezing and Thawing of Ultra-high Molecular Weight of Polystyrene in p-xylene.

Number of Freezing Cycles	Polymer Con. g/l	$\overline{M}_w \cdot 10^6$	$\overline{M}_n \cdot 10^6$	$\overline{M}_w/\overline{M}_n$	Scissions per Original Molecule
0	–	7.29	6.45	1.1	–
15	2.5	4.67	2.69	1.7	1.4
30	2.5	3.47	2.03	1.7	2.2
45	2.5	2.60	1.45	1.8	3.5
15	0.15	4.86	1.68	2.9	2.9
30	0.15	3.93	1.15	3.4	4.6
45	0.15	3.12	0.985	3.2	5.5

degradation is quite high, the MWD had not been transformed into the most probable distribution ($\overline{M}_w/\overline{M}_n$ = 2). As may be seen from Table 1, polydispersity changed in quite a different way at lower polystyrene concentrations (0.15 g/l). The results indicate a difference between the degradation mechanisms at high and low concentrations. Obtaining an H of 2.0 does not necessarily mean that the process follows random scission kinetics[10]. A nonrandom scission followed by a random hydrogen abstraction from adjacent polymer chains leads to secondary radicals, which may undergo β-scission. In such a case H approaches two, although the primary degradation reaction was nonrandom.

An elucidative paper concerning the problems in finding valid criteria for random degradation has been published by Scott[10]. He concluded that, in most cases, the use of the relative decreases in $\overline{M}_w$ and $\overline{M}_n$ with scission provides the best criterion from random scission. This procedure was used to evaluate the degradation reactions occurring during repetitive freezing and thawing of polystyrene solutions in p-xylene[9]. Polymer concentrations between 2.5 - 20 g/l were investigated. The results are shown in Figure 2. Further data on this system are given below. This type of plot utilizes the

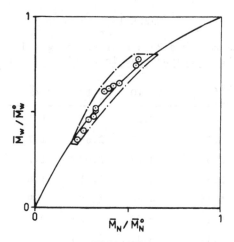

FIGURE 2

Relative Changes in $\overline{M}_w$ Versus Relative Changes in $\overline{M}_n$. Data for Cryogenic Degradation of Polystyrene Solutions (2.5 - 20 g/l in p-xylene). The Solid Line Corresponds to a Random Scission Mechanism.

maximum information available and includes, in essence, the increase in chain scission as well as the changes in H. The solid line in Figure 2 is derived from equations (4) and (5):

$$\overline{M}_n/\overline{M}_n^o = 1/ (1 + \alpha) \qquad\qquad (4)$$

$$\frac{\overline{M}_w}{\overline{M}_n^o} = \frac{2}{\alpha H}\left\{1 - \frac{1}{\alpha}\left[1 - \int_0^\infty \frac{W(y)}{mn/\overline{M}_n^o} \cdot e^{-\beta} \, dy\right]\right\} \qquad (5)$$

assuming $\overline{M}_n^o \gg 100$ and $\beta \ll 1$.

$\overline{M}_n$ and $\overline{M}_w$ are, respectively, the number and weight average molecular weight after scission, $\overline{M}_n^o$ and $\overline{M}_w^o$ are the corresponding values before scission, α is the number of scissions per initial number average molecular weight molecule, W (y) is the weight fraction of y-mers initially present, β is the fraction of the bonds (between monomer units) that have undergone scission, and m is the molecular weight of the monomer. Equation (5) can be evaluated for several common distribution functions, but for values of H close to

unity the results are nearly independent of the type of MWD assumed.
The solid line in Figure 2 was calculated for H = 1.13, which is the
heterogeneity index of the starting polymer. Data for polystyrene
solutions with concentrations $\geq$ 2.5 g/l are in excellent agreement
with the calculation. This analysis requires carefully determined
molecular weight averages. The effect of inaccuracies in these
measurement are shown in Figure 2. The area indicated in the dia-
gram describes the deviations that occur assuming a maximum error in
$\overline{M}_n$ and $\overline{M}_w$ of $\pm$ 5% and $\pm$ 3%, respectively. The left hand side of the
area corresponds to the case when $\overline{M}_n$ is 5% below and $\overline{M}_w$ 3% above
the true values. The right hand side boundary corresponds to the
case when $\overline{M}_n$ is 5% above and $\overline{M}_w$ 3% below the true values. We have
assumed that $\overline{M}_n$ and $\overline{M}_w$ of the original sample were correctly determined. As may be seen from Figure 2, the accuracy in $\overline{M}_n$ and $\overline{M}_w$ was
sufficiently high provided that $\overline{M}_n$ was not notoriously underestimated when $\overline{M}_w$ was overestimated and vice versa.

The above mentioned procedure was also employed to elucidate
the degradation reactions occurring during recycling of polycarbonate in a capillary rheometer at high shear stresses[11,12]. It has
been claimed in literature that the degradation of polycarbonate
during repeated injection molding is a random scission process[13].
However, Scott recalculated some of the earlier data and found nonrandom characteristics[10]. To analyze our data according to Scott,
we thus plotted the relative changes in $\overline{M}_w$ versus the relative
changes in $\overline{M}_n$ and the result was compared with that expected from
theory. As previously mentioned, the theoretical treatment requires
that the original MWD can be described by a distribution function.
David and Baeyens-Volant have proposed several criteria that may be
used to verify if a given MWD obeys a specific distribution function
[14]. For the Schulz-Zimm distribution (generalized Poisson distri-
bution) it was found that $(\overline{M}_n + \overline{M}_z)/\overline{M}_w = 2$ and $1 < \overline{M}_z/\overline{M}_w < 2$. In
our case $(\overline{M}_n + \overline{M}_z)/\overline{M}_w = 2.1$ and $\overline{M}_z/\overline{M}_w = 1.65$. The MWD or virgin
polycarbonate may thus very well be approximated with a Schulz-Zimm
distribution. It was found that all experimental results were locat-

ed significantly above the theoretical curve, which implies that
polycarbonate did not degrade according to a random scission
mechanism under these conditions[11,12]. Unfortunately, this analysis does not give any indication of the character of the operating
scission mechanism. To further elucidate the nature of this reaction, an even more sophisticated analysis of the degradation was
performed (see below).

SCISSION INTENSITY DETERMINATIONS

Numerical model for chain scission

A detailed analysis of chain scission kinetics may be carried
out according to a procedure described by Fukutomi et al.[15]. It
is based on a general equation for scission reactions proposed by
Simha[16]. Let k_j^i be the scission rate constant for the formation
of a j-mer from an i-mer (j < i) and let m be the highest degree of
polymerization, which exists in the reaction system (n_j (t) $\equiv$ 0 for
j > m). n_i (t) is the number of i-mers at time t. The following set
of differential equations may then be derived[16].

$$\frac{d\,n_m\,(t)}{dt} = \sum_{j=1}^{m-1} k_j^m\,n_m\,(t)$$

$$\vdots$$

$$\frac{d\,n_j\,(t)}{dt} = \sum_{i=j+1}^{m} (k_j^i + k_{i-j}^i)\,n_i\,(t) - \sum_{i=1}^{j-1} k_i^j\,n_j\,(t)$$

$$\vdots$$

$$\frac{d\,n_1\,(t)}{dt} = \sum_{i=2}^{m} (k_1^i + K_{i-1}^i)\,n_1\,(t) \tag{6}$$

All repetitive units in the polymer are assumed to be identical.
As a consequence of the existing symmetry, $k_j^i = k_{i-j}^i$, equation (6)
may be written in a simplified form:

$$\frac{d\,n_m\,(t)}{dt} = 2 \sum_{i=j+1}^{m} k_j^i\,n_i\,(t) - \sum_{i=1}^{j-1} k_i^j\,n_j\,(t) \tag{7}$$

As the degree of polymerization often is high it may neither be numerically or experimentally feasible to distinguish a j-mer from a (j+1)-mer. It is thus reasonable to divide the molecular weight distribution (MWD) into n intervals and state that a molecule with a molecular weight ν belongs to the interval I_i if (i-1_w $\nu \le$ iw (w chosen arbitrarily). K_j^i now designates the scission rate constant for molecules in interval I_i to form molecules belonging to I_j (j < i). N_j (t) is the number of molecules in interval I_j at time t. Equation 6 may then be written as follows:

$$\left(\frac{dN_n}{dt}\right)_{t=t_k} = - \sum_{i=1}^{n-1} K_i^n\,N_n\,(t_k)$$

$$\vdots$$

$$\left(\frac{dN_j}{dt}\right)_{t=t_k} = 2 \sum_{i=j+1}^{n} K_j^i\,N_i\,(t_k) - \sum_{i=1}^{j-1} K_i^j\,N_j\,(t_k)$$

$$\vdots$$

$$\left(\frac{dN_1}{dt}\right)_{t=t_k} = 2 \sum_{i=2}^{n} K_1^i\,N_i\,(t_k) \tag{8}$$

K = 1, 2, 3, ..., m.

Numerical values of N_j (t_k) and $\left(\frac{dN_j}{dt}\right)_{t=t_k}$ may be derived from the changes in MWD. The number of unknowns, K_j^i, (2 $\le$ i $\le$ n; 1 $\le$ j $\le$ i - 1) are n(n-1)/2. In the case of symmetry, $K_j^i = K_{i-j}^i$, the number of unknowns are reduced to:

$$\frac{n\,(n-1)}{4} + \frac{1}{2} \left[\frac{n}{2}\right] \approx \frac{n^2}{4} \tag{9}$$

As indicated above m designates the number of MWD's (GPC traces). The total number of equations in Equation (8) then amounts to n . m.

As the number of unknowns has to be lower than the number of equations we obtain (provided that $K_j^i = K_{k-j}^i$):

$$\frac{n^2}{4} \leq n \cdot m \text{ or } n \leq 4 m \tag{10}$$

From Equation (10) it is evident that the number of intervals cannot be chosen arbitrarily high but depends on the number of MWD's (determined at t_1, ..., t_m). It is also important to note that if n is close to its upper limit, numerical instability occurs. On the other hand if n is too small we obtain too coarse a model. $2 m \leq n \leq 3 m$ was found to be a reasonable compromise. The overdetermined system of equations is then solved by the method of least squares.

Cryogenic degradation of polystyrene solutions

Solutions of ultrahigh molecular weight polystyrene ($\overline{M}_w = 7.3 \cdot 10^6$) in p-xylene were frozen in liquid nitrogen[9]. In total, the polymer was exposed to 45 freezing cycles. GPC analyses were performed after 0, 15, 30, and 45 cycles. Two different degradation mechanisms were distinguished at low and at high concentrations. The change between mechanisms took place between 1.0 and 2.5 g/l. MWD's obtained for polystyrene concentrations of 2.5 g/l and 0.15 g/l are shown in Figures 3 and 4, respectively.

The MWD's were divided into 10 intervals. The interval size (= w) was chosen so that less than 0.5% of the existing molecules exceeded a weight of 10 w. We then obtain the following intervals:

$$I_1 = (0 - 1.5 \cdot 10^6)$$
$$I_2 = (1.5 \cdot 10^6 - 3.0 \cdot 10^6)$$
$$\vdots$$
$$I_{10} = (13.5 \cdot 10^6 - 15 \cdot 10^6)$$

The MWD's were normalized and the determination of $N_j(t_k)$ and $(\frac{dN_j}{dt})_{t=t_k}$ were performed as follows: Every interval was subdivided

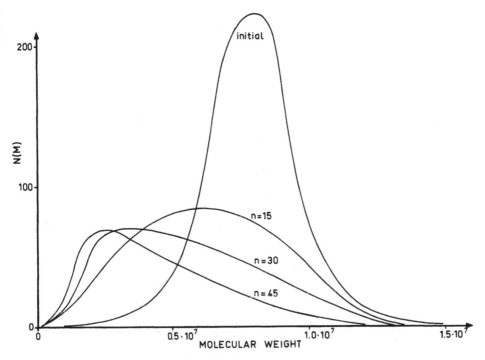

FIGURE 3
Changes in MWD as a Function of Freezing Cycles for 2.5 g/l Solutions
of Polystyrene in p-xylene.

into five equal parts and the frequency at each point was divided
by the corresponding molecular weight. These values were then used
for the integration (by Simpson's formula) over every original interval. A measure proportional to the number of molecules in each
interval, N_j (t_k), was thus obtained. This procedure was repeated
for all MWD's. The derivatives $(\frac{dN_j}{dt})_{t=t_k}$, were determined by
graphical as well as numerical derivatization of $N_j(t_k)$. The relative
changes in the number of molecules (k · N) in each interval during
the repetitive freezing and thawing were determined.

Figures 5 and 6 show the result for a polystyrene concentration of 2.5 g/l. Values of K_j^i were then plotted as a function of
molecular weight difference between the chain segments formed during

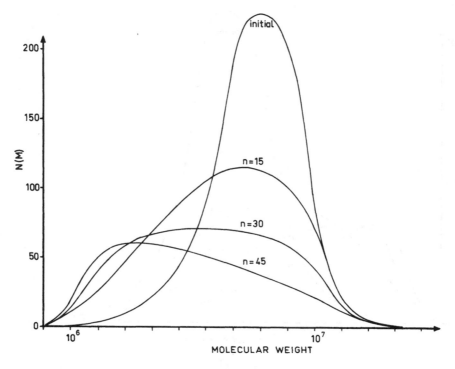

FIGURE 4

Changes in MWD as a Function of Freezing Cycles for 0.15 g/l Solutions of Polystyrene in p-xylene.

scission. As may be seen from Figure 7 scission rate constants were independent of chain length as well as of the macromolecule. This means that the degradation reaction followed random scission kinetics (K_j^i was constant).

A corresponding data treatment was performed for a polystyrene concentration of 0.15 g/l. The relative changes in the number of molecules in each interval during the experiment are shown in Figures 8 and 9. Very large variations in K_j^i were now observed including negative chain scission. This is probably the result of secondary reactions and it might be interpreted as a combination of the polymer radicals formed. Further experiments are needed to elucidate the exact nature of these secondary reactions.

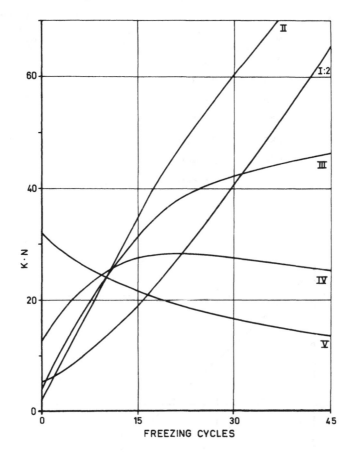

FIGURE 5

The Relative Number of Macromolecules in Intervals 1-5 as a Function
of Freezing Cycles. Data for 2.5 g/l Solutions of Polystyrene in
p-xylene.

Shear degradation of polycarbonate

The above mentioned procedure was also used for the degradation
analysis of polycarbonate, which was repeatedly extruded in a capillary rheometer[11,12]. The extrusion was carried out at 320°C with
a shear stress of 0.15 MPa. After each cycle samples were removed
for GPC analysis (see Table 2).

TABLE 2

Repetitive Extrusion of Polycarbonate at 320°C and τ_w = 0.15 MPa.

Number of Extrusions	$\overline{M}_N$	$\overline{M}_W$	$\overline{M}_W/\overline{M}_N$	Number of Scissions Per Molecule
0	13,780	30,200	2.19	–
2	12,700	28,300	2.23	0.08
3	11,100	25,800	2.31	0.25
4	9,430	22,000	2.34	0.46
5	9,080	21,200	2.34	0.50
6	7,950	19,400	2.44	0.74
7	6,800	17,600	2.59	1.00

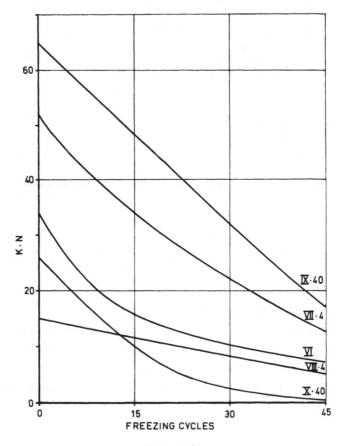

FIGURE 6

The Relative Number of Macromolecules in Intervals 6-10 as a Function of Freezing Cycles. Data for 2.5 g/l Solutions of Polystyrene in p-xylene.

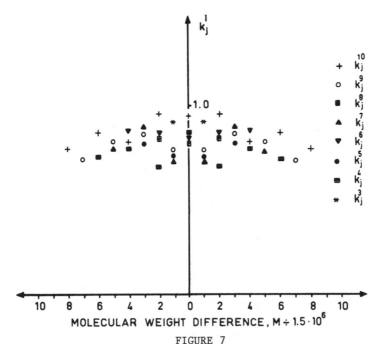

FIGURE 7

Scission Rate Constants Plotted Versus the Molecular Weight Difference Between Segments Formed. Data for 2.5 g/1 Solutions of Polystyrene in p-xylene.

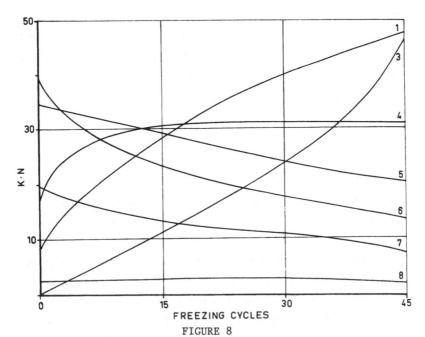

FIGURE 8

The Relative Number of Macromolecules in Intervals 1 and 3-8 as a Function of Freezing Cycles. Data for 0.15 g/1 Solutions of Polystyrene in p-xylene.

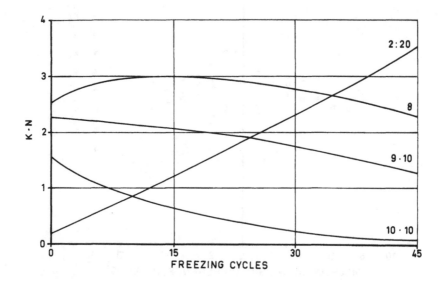

FIGURE 9

The Relative Number of Macromolecules in Intervals 2 and 8-10 as a Function of Freezing Cycles. Data for 0.15 g/l Solutions of Polystyrene in p-xylene.

The MWD's were divided into 15 intervals. The interval size (= w) was determined so that less than 0.5% of the longer molecules were excluded (>15w). In this case a w-value of 10,000 was chosen, which resulted in the following intervals:

I_1 = (0 - 10,000)
I_2 = (10,000 - 20,000)
.
.
.
I_{15} = (140,000 - 150,000)

The MWD's were then normalized and the number of molecules in each interval was determined as described above. The resulting K_j^i-values are shown in Figure 11. The abscissa shows the molecular weight difference between the fragments after scission. Independent of polymer chain length there was a clear preference for scissions to occur close to the middle of the macromolecules. It was also obvious that the longer molecules were more susceptible to scission than the shorter ones. Our results comply with a theory, which pre-

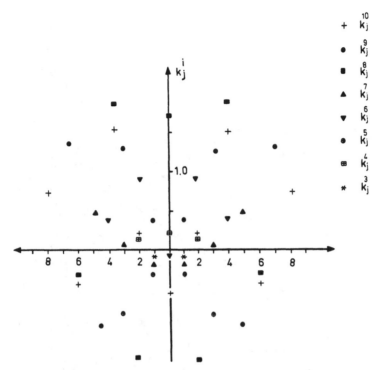

MOLECULAR WEIGHT DIFFERENCE, $M \div 1.5 \ 10^6$

FIGURE 10

Scission Rate Constants Plotted Versus the Molecular Weight Difference Between Segments Formed. Data for 0.15 g/l Solutions of Polystyrene in p-xylene.

dicts a Guassian distribution for the location of rupture centered around the middle of the polymer chain[7,8].

Error analysis

      Possible errors in the scission intensity analysis are discussed below:

             1. The mathematical model is based on the assumption that all
             macromolecules are linear and that no recombination or cross-
             linking occurs. The presence of such reactions means that K_j^i-
             values lose their identity as scission rate constants.
             2. The numerical model is not quite correct. By dividing the
             MWD's in intervals and assuming that all molecules in each

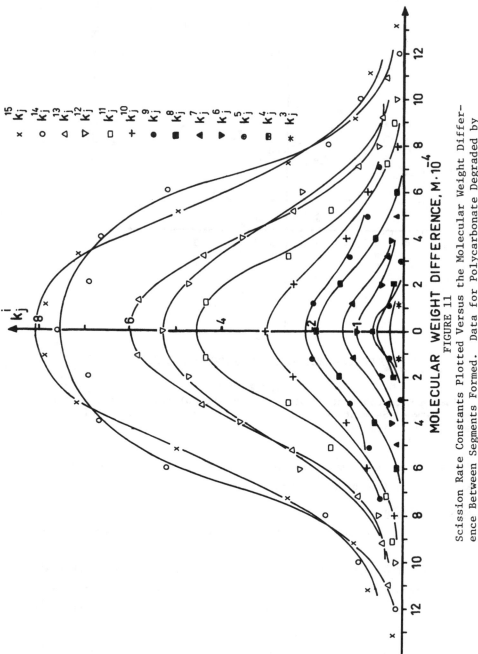

MOLECULAR WEIGHT DIFFERENCE, M·10⁻⁴
FIGURE 11

Scission Rate Constants Plotted Versus the Molecular Weight Difference Between Segments Formed. Data for Polycarbonate Degraded by
Repetitive Extrusion in a Capillary Rheometer.

interval have the same scission intensity, we are making an
approximation. The error may be decreased by dividing the
MWD´s in smaller intervals. However, numerical instability
occured when the number of intervals was larger than 3 m,
where m is the number of MWD´s. This means that the accuracy
may be improved by determining additional MWD´s at different
stages of the degradation process.

3. Inaccuracy in MWD´s. The MWD´s are preferably determined
by gel permeation chromatography (GPC), which gives the most
accurate results. The largest errors are observed at the
ends of the distributions, where there might be difficulties
in measuring the low intensities. GPC-traces exhibiting good
base-lines are always required to minimize such errors.

4. Approximation errors. This kind of errors occur mainly
as a result of three calculations: (a) the area determination
by Simson's formula, (b) the graphical and numerical derivation, and (c) the least squares approximation. By using 5
digits throughout the calculations, these errors were considered to be negligable.

ACKNOWLEDGEMENT

The author would like to thank Mr. B. Tarras-Wahlberg for his
valuable help in performing the scission intensity determinations.

REFERENCES

1. Abbås, K.B., Appl. Polym. Symp., in press.

2. Abbås, K.B., Appl. Polym. Sci., 19, 2991 (1975).

3. Drott, E.E., and Mendelson, R.A., J. Polym. Sci., A-2, 8, 1361
   (1970).

4. Zimm, B.H., and Kilb, R.W., J. Polym. Sci., 37, 19 (1959).

5. Abbås, K.B., and Sorvik, E.M., J. Appl. Polym. Sci., 17, 3577
   (1973).

6. Bueche, F., J. Appl. Polym. Sci., 4, 101 (1960).

7. Glynn, P.A.R., Van der Hoff, B.M.E., and Reilly, P.M., J.
   Macromol. Sci.-Chem., A-6, 1653 (1972).

8. Abdel-Alim, A.H., and Hamielec, A.E., J. Appl. Polym. Sci., <u>17</u>, 3769 (1973).

9. Abbås, K.B., and Porter, R.S., J. Polym. Sci., Chem. Ed., <u>14</u>, 553 (1976).

10. Scott, K.W., J. Polym. Sci., <u>C 46</u>, 321 (1974).

11. Abbås, K.B., <u>Proceedings of the 2nd International Symposium on Degradation and Stabilization of Polymers</u>, Dubrovnik, Yugoslavia, 1978.

12. Abbås, K.B., Polym. Eng. Sci., in press.

13. Glockner, G., Plast. Kaut., <u>15</u>, 632 (1968).

14. David, C., and Baeyens-Volant, D., Eur. Polym. J., <u>14</u>, 29 (1978).

15. Fukutomi, T., Tsukada, M., Kakurai, T., and Naguchi, T., Polym. J., <u>3</u>, 717 (1972).

16. Simha, R., J. Appl. Phys., <u>12</u>, 569 (1941).

CHARACTERIZATION OF POLYMERS WITH LONG CHAIN BRANCHING-DEVELOPMENT OF THE MOLECULAR WEIGHT AND BRANCHING DISTRIBUTION (MWBD) METHOD

G.N. Foster   T.B. MacRury

Union Carbide Corporation   Union Carbide Corporation
Bound Brook, New Jersey   South Charleston, West Virginia

A.E. Hamielec

Department of Chemical Engineering
McMaster University
Hamilton, Ontario

ABSTRACT

A molecular weight and branching distribution (MWBD) method is presented for measuring long chain branching (LCB) as a function of molecular weight in such polymers as polyvinyl acetate and polyethylene. The method requires intrinsic viscosity (IV) and size exclusion chromatographic (SEC) data on the polymer, and a polystyrene calibration determined under the same solvent-temperature conditions. The MWBD method expresses the size separation with elution volume (V) as $\ln ([\eta]_i M_{N,i}) = f(V_i)$, and assumes IV to be described by a polynomial of the form - $\ln [\eta]_i = \ln K + a \ln M_{N,i} + b (\ln M_{N,i})^2 + C(\ln M_{N,i})^3$. The number-average molecular weight and the number of branch points on a molecular basis are obtained as a function of elution volume. From this, integration across the chromatogram provides the number average molecular weight $(\overline{M}_N)$ and the number-average number of branch points per molecule and per 1000 carbon atoms $(\overline{B}_N$ and $\overline{\lambda}_N$, respectively). As part of the MWBD method, C-13 nuclear magnetic resonance spectroscopy is used to validate the method and the theoretical assumptions made. The utility of the MWBD method is illustrated using a series of commercial high-pressure, low density polyethylenes and polyvinyl acetates. Correlation of the LCB results are made to rheological and blown film properties and to expectations based on reaction kinetics.

INTRODUCTION

Free radical polymerization leads to long chain branching (LCB) in commercial polymers such as polyvinyl acetate (PVAc) and low density polyethylene (LDPE) via a chain transfer to polymer mechanism[1]. Random branch lengths and random spacings between branch points are most probable for such polymers. For LDPE's made by the high-pressure (HP) tubular reactor process, polymerization temperatures change significantly through the reaction zones. Thus, these polymers can be an even more complex mixture of branched chains, varying in the number of LCB points per molecule with molecular weight (MW) or the degree of polymerization for the molecular species. LCB content in commercial PVAc and HP-LDPE's ranges from 1 to 20 branch points per molecule. More specifically for HP-LDPE's, the LCB frequency-another useful parameter-will vary from 0.5 to 5 LCB points per 1000 carbon atoms. Short chain branching frequency (SCB) in HP-LDPE's near a 0.920 gm/cc density will range between 10 to 20 branch points per 1000 carbon atoms depending on SCB length, i.e., methyl, ethyl, etc. Typically, the butyl branch length predominates, with minor amounts of pentyl branches. Both are formed by a cyclic intramolecular chain transfer mechanism called "back biting."[1] Varying amounts of other branch lengths are possible such as methyl, gem-dimethyl, or ethyl depending on the chain transfer agent used.

LCB has a pronounced effect on the flow behavior of polymers under shear and extensional flow. Increasing LCB will increase melt viscosity-shear rate sensitivity and elasticity.[2] Under melt extensional deformation, LCB promotes strain hardening or increased extensional viscosity and melt strength characteristics. This, in turn, diminishes the ability to draw-down polymer melts to thin cross-sections.[3] Solid state properties involved with stress or impact failures are dependent on the macrocrystalline structure formed under stress or orientation conditions which, in turn, will be controlled by LCB and molecular size through intercrystalline tie linking. Thus, environmental stress cracking and low-tempera-

ture brittleness can be strongly influenced by the LCB. In HP-LDPE's
SCB affects microcrystalline-related properties such as density, heat
of fusion, stiffness (secant modulus)[4] but does not influence melt
shear or extensional flow behavior as does LCB. Therefore, the
ability to measure LCB and its MW distribution and to tailor product
performance by controlling LCB is most important.

Such LCB polymers as described confound the molecular weight
analysis by conventional gel permeation chromatography (GPC) or, in
a broader sense, by size exclusion chromatography (SEC), which en-
compasses GPC using cross-linked polystyrene packed columns as well
as controlled pore size silica packed columns. SEC separates molecules according to hydrodynamic size. The chromatographic separation according to size is directly related to the product of intrinsic viscosity, $[\eta]$, and MW as shown by Benoit et al.[5] Recent
publication by Hamielec and Ouano[6] has shown that the instantaneous MW in the SEC detector cell is more correctly a number
average MW, $M_N(V)$. Therefore, the molecular size at a given time
or, for a particular elution volume, in the detector cell is equal to
the $[\eta] \cdot M_N(V)$ product. Hence, the detector cell—whether within a
differential refractometer or an infrared spectrophotmeter—will
contain a complex mixture of molecules having the same hydrodynamic
size, $[\eta] \cdot M_N(V)$, but different MW's due to LCB variation.

Drott[7], as well as Ram and Miltz[8], used SEC (conventional GPC)
methods for correcting chromatographic results for LCB. Both inter-
pretations ignored MW dispersion in the detector cell. The Drott
method assumed that LCB frequency was not a function of MW—this has
been recognized to be in error. Both methods were for correcting
the $\overline{M}_N$ and the higher MW moments for LCB. As such, they did not
provide information as to how LCB may be distributed with MW. In
principle, the limitations associated with these early methods
could be overcome if on-line detectors for absolute $M_N(V)$ and weight
average MW $M_W(V)$ measurements were available. Low angle laser light
scattering photometry can provide for $M_W(V)$ detection at elevated
temperatures. However, solution viscosity detectors for $M_N(V)$

measurement are not yet developed that can operate at 130 to 140°C
for polyolefin analysis and, in addition, can be used with the small
elution volumes associated with present, high-speed SEC equipment.

Herein is presented a new method for SEC interpretation which
accounts for the instantaneous polydispersity due to LCB within the
detector cell and which allows for the calculations of LCB content
and frequency across the chromatogram as a function of MW. Hereafter,
this method will be referred to as the molecular weight-branching
distribution (MWBD) method. Applications for the MWBD method will
be highlighted using selected commercial HP-LDPE's and specially
synthesized PVAc's. Further, an absolute method now exists for
measuring LCB frequency on whole or fractionated polymers using
C-13 nuclear magnetic resonance (NMR) spectroscopy. C-13 NMR spectroscopy has been used to measure LCB in HP-LDPE's as well as for
SCB measurements. We will show how it can be used to check the
validity of the MWBD method.

DISCUSSION

Theory

As pointed out by Hamielec and Ouano,[6] the separation of
a branched polymer by size in the SEC process results in molecular
species of different molecular weights eluting at the same volume.
Thus, the molecular weight of these species $M(V)$ is not monodispersed. Therefore, the universal calibration procedure[5] used for
linear polymers is given by

$$\ln J(V) = A + BV + \ldots, \tag{1}$$

where the hydrodynamic volume $J(V)$ is defined as

$$J(V) = [\eta](V)M(V) \tag{2}$$

and $A, B, \ldots$ are the coefficients determined from fitting $\ln J(V)$ as
a function of V.

Using the fact that at volume V all species have the same
hydrodynamic size, we can write

$$J_{ps}(V) = J_1(V) = J_2(V) = \ldots, \tag{3}$$

where species 1,2,... may represent linear, branched, copolymers, or blends of homopolymers and "ps" represents polystyrene or some other linear narrow distribution standards used for calibration. Now the intrinsic viscosity in the detector cell can be computed from

$$[\eta](V) = W_1[\eta]_1 + W_2[\eta] + \ldots, \tag{4}$$

where W_1, W_2,... are the weight fractions of the species and $[\eta]_1$, $[\eta]_2$,...are the corresponding intrinsic viscosities. This expression can also be written as

$$[\eta](V) = J_1 \frac{W_1}{M_1} + J_2 \frac{W_2}{M_2} + \ldots, \tag{5}$$

and using equation (3) leads to

$$[\eta](V) = J_{ps}(V)\left[\frac{W_1}{M_1} + \frac{W_2}{M_2} + \ldots\right] \tag{6}$$

However, since the instantaneous number-average molecular weight is defined as

$$M_N(V) = \frac{M_1}{W_1} + \frac{M_2}{W_2} + \ldots, \tag{7}$$

equation (6) can be rewritten in the following form:

$$[\eta](V) M_N(V) = [\eta]_{ps}(V)M_{ps}(V) \tag{8}$$

Thus the molecular weight given by the universal calibration procedure is M_N and not M_W as has been suggested in the past.[5,7,9,10] The choice of M_W as the proper molecular weight average to couple with the intrinsic viscosity for the universal calibration procedure perhaps evolved from the intuitive feeling that for most polymers the weight average molecular weight is the closest average to the viscosity average molecular weight. Note that for linear polymers

$M_N(V) = M_W(V) = M_Z(V)$ since the contents of the detector cell are
monodispersed.

The best method for calculating $M_N(V)$ across the chromatogram
would be to use an on-line viscometer to measure $[\eta](V)$ directly,
and then to determine $M_N(V)$ from

$$M_N(V) = J(V)/[\eta](V) \qquad\qquad (9)$$

where $J(V)$ is obtained from equation (1)

$$J(V) = \exp(A + BV + \ldots) \qquad\qquad (10)$$

Unfortunately, an on-line viscometer which can provide instantaneous
$[\eta]$ values at high and low temperatures is not available. Although
batch viscometers have been used in the past,[11] the use of high-
speed SEC makes them useless due to the small elution volumes.
Ouano[12] performed on-line measurements of $[\eta]$ at low temperatures
but not at high temperatures.

We therefore propose an indirect method for obtaining the
variation of the intrinsic viscosity and number average molecular
weight across the chromatogram. Let us express the intrinsic vis-
cosity-molecular weight relationship for a polymer with long chain
branching (LCB) in a form similar to that used by Ram and Miltz,[8]
namely

$$\ln\left([\eta](V)\right) = \ln K + a\ln M_N(V) + b\left(\ln M_N(V)\right)^2 + c\left(\ln M_N(V)\right)^3 \qquad (11)$$

where K and a are the Mark-Houwink constants for the linear homo-
logue in the same solvent and at the same temperature as the SEC
measurements. If it is then assumed that there is no LCB below a
certain molecular weight M_L, the parameter c in equation (11) can
be replaced with

$$c = -b/\ln M_L \qquad\qquad (12)$$

The value normally assumed for M_L is between 5000 and 10,000.[8,13]

We are now left with one unknown b which can be determined
from the measured whole polymer intrinsic viscosity in the following
manner. First, one estimates a value for b, and calculates $[\eta](V)$
and $M_N(V)$ across the chromatogram using the universal calibration

curve. The whole polymer intrinsic viscosity is obtained from

$$[\bar{\eta}] = \int F(V)[\eta](V) \, dV \tag{13}$$

where $F(V)$ is the normalized concentration detector response. The calculated $[\eta]$ value using equation (13) is compared with the measured value. Further values of b are then tried until the difference between the measured and calculated intrinsic viscosities are minimized.

Once b has been determined, the true whole polymer number average molecular weight and higher averages can be calculated as follows:

$$\bar{M}_N = \left\{ \int \frac{F(V)}{M_N(V)} \cdot dV \right\}^{-1} \tag{14}$$

$$\bar{M}_N = \int F(V)M_N(V) \, dV, \tag{15}$$

$$\bar{M}_Z = \int F(V) \left\{ M_N(V) \right\}^2 \, dV \bigg/ \int F(V)M_N(V) \, dV \tag{16}$$

It should be pointed out that, whereas the number-average molecular weight is correct, the weight-average, Z-average, and other higher moments are only approximate and less than the true values. This is a consequence of the universal calibration method giving the instantaneous number-average molecular weight across the chromatogram.

The instantaneous weight-average molecular weight $M_W(V)$ can also be measured by using an on-line, low angle laser light scattering device. Such measurements have been performed successfully both at low temperatures[14,15] and high temperatures.[16] When combined with the method presented here for determining $M_N(V)$, the polydispersity in the detector cell

$$(V) = M_W(V)/M_N(V) \tag{17}$$

could be calculated across the chromatogram. Work on the combination
of these two methods will be repored at a later date.

Since our indirect method produces both the linear and branched
intrinsic viscosities across the chromatogram, it is possible to
estimate several LCB parameters, not only of the whole polymer, but
also as a function of elution volume or number-average molecular
weight. The branching factor $G(V)$ can be written as

$$G(V) = \left\{ [\eta]_b(V)/[\eta]_\ell(V) \right\}^{1/\varepsilon} \tag{18}$$

where $[\eta]_b(V)$ is the instantaneous branched intrinsic viscosity,
$[\eta]_\ell(V)$ is the instantaneous linear intrinsic viscosity, and ε is a
constant to be determined. For star polymers a value of $\varepsilon = 0.5$ has
been obtained[17,18] and studies[19] of model comb polymers indicate a value of 1.5. Other work[20] has suggested that, at low LCB
frequencies, ε is near 0.5. One might expect, for a random LCB conformation of higher branch frequency, that ε will lie between 0.7
and 1.3, i.e. somewhere between a comb and a star configuration.

Zimm and Stockmayer[21] have derived a theoretical relationship between $G(V)$ and the number-average number of LCB points per
molecule $B_N(V)$, namely

$$G(V) = \left[\left(1 + \frac{B_N(V)}{7} \right)^{1/2} + \frac{4B_N(V)}{9} \right]^{-1/2} \tag{19}$$

Thus, by assuming some appropriate value for ε, the number average
number of LCB points per molecule can be calculated as a function
of the elution volume from equations (18) and (19). To obtain the
number-average number of LCB points per molecule for the whole polymer, the following equation can be used:

$$\bar{B}_N = \int \frac{B_N(V)F(V)}{M_N(V)} \cdot dV \bigg/ \int \frac{F(V)}{M_N(V)} \cdot dV \tag{20}$$

$$= \bar{M}_N \int \frac{B_N(V)F(V)}{M_N(V)} \cdot dv \qquad (21)$$

Another parameter commonly used for characterizing long chain branching is the branching frequency per 500 monomeric repeat units or the number-average number of branch points per 500 monomeric repeat units or more commomly expressed per 1000 carbon atoms for polyethylenes. It can be calculated across the chromatogram from

$$\bar{\lambda}_N(V) = 500 \, M_R \left(B_N(V)/M_N(V) \right) \qquad (22)$$

were M_R is the molecular weight of the monomeric repeat unit. The number-average number of branch points per 500 repeat units for the whole polymer is

$$\bar{\lambda}_N = 500 \, M_R \int \frac{B_N(V)F(V)}{M_N(V)} \, dv \qquad (23)$$

or from equation (21)

$$\bar{\lambda}_N = 500 \, M_R \frac{\bar{B}_N}{\bar{M}_N} \qquad (24)$$

In an analogous fashion, the weight-average number of branch points per molecule (B_W) and per 500 monomeric repeat units (λ_W) can be calculated across the chromatogram and for the whole polymer. First, $B_W(V)$ can be determined from the Zimm-Stockmayer equation:[21]

$$G(V) = \frac{6}{B_W(V)} \left[\frac{1}{2} \frac{\left[2+B_W(V)\right]^{\frac{1}{2}}}{\left[B_W(V)\right]^{\frac{1}{2}}} \ln\left\{ \frac{\left[2 + B_W(V)\right]^{\frac{1}{2}} + \left[B_W(V)\right]^{\frac{1}{2}}}{\left[2 + B_W(V)\right]^{\frac{1}{2}} - \left[B_W(V)\right]^{\frac{1}{2}}} \right\} - 1 \right] \qquad (25)$$

Then $\bar{B}_W$, $\lambda_W(V)$, and $\bar{\lambda}_W$ can be found using the following equations:

$$\bar{B}_W = \int \frac{B_W(V)F(V)}{M_N(V)} \cdot dV \left/ \int \frac{F(V)}{M_N(V)} \cdot dV \right. \qquad (26)$$

$$\lambda_W(V) = 500\ M_R \left(B_W(V)/M_N(V) \right) \tag{27}$$

and

$$\overline{\lambda}_W = 500\ M_R\ \frac{\overline{B}_W}{\overline{M}_N} \tag{28}$$

It should be noted that, in order to calculate the weight average number of LCB points per 500 monomeric repeat units, one must divide $\overline{B}_W$ by $\overline{M}_N$. The use of $\overline{M}_W$ here is incorrect.

This method for characterizing branched polymers will here-after be referred to as the molecular weight and branching distri-bution (MWBD) method. In the following sections it will be applied to some branched PVAc's and HP-LDPE's.

Polyvinyl Acetate Studies

Table 1 summarizes the property data for the first two poly vinyl acetate samples used to illustrate the utility of the MWBD method. These samples, PVAc-Lot 1 and PVAc-Lot 3, are commercial standards supplied by Aldrich Chemical Co. The third sample, PVAc-E4 was synthesized under kinetically controlled conditions (isothermal, T=60°C [AlBN] = 10^{-5} gmole/ℓ conversion level of 48.5 percent). Under conditions of low radical initiation, Graessley[22] has shown that the following set of equations describes molecular weight and branching development in the bulk polymerization of vinyl acetate.

$$\frac{dQ_0}{dX} = C_M - \frac{KQ_0}{(1-X)} \tag{29}$$

$$\frac{dQ_1}{dX} = 1 \tag{30}$$

$$\frac{dQ_2}{dX} = 2 \left[1 + \frac{KX}{1-X} \right] \left[\frac{1 + \dfrac{C_p Q_2}{1-X} + \dfrac{KX}{1-X}}{C_M + \dfrac{C_p X}{1-X}} \right] \tag{31}$$

$$\frac{d(Q_0 \bar{B}_N)}{dX} = \frac{C_p X + KQ_0}{(1-X)} \tag{32}$$

where

$$C_M = \frac{k_{fm}}{k_p} \quad ; \quad C_p = \frac{k_{fp}}{k_p} \quad ; \quad K = \frac{k_p^*}{k_p}$$

$$\bar{M}_N = M_m Q_1 \Big/ Q_0 \qquad \bar{M}_W = M_m Q_2 \Big/ Q_1$$

and k_{fm} is the transfer to monomer constant.

k_p is the propagation constant.

k_{fp} is the transfer to polymer constant.

k_p^* is the constant for terminal double bond reactions.

M_m is the monomer molecular weight.

Q_0, Q_1, Q_2, are the zeroeth, first and second moments of the molecular weight distribution.

$\bar{B}_N$ is the number-average number of branch points per molecule.
A solution of equations (29) to (32) is shown plotted in Figure 1 for a set of the kinetic parameters C_M, C_p, and K. LCB development increases at higher conversions raising $\bar{M}_W$ and $\bar{B}_N$ significantly. Solutions for various sets of these parameters are shown tabulated in Table 2 for conversions of 50 and 60 percent. It is expected that the average number of branch points per molecule for the whole polymer (Sample E-4) might be in the range shown, 0.6-2.8 branches per molecule. Graessley[22] states that literature values range from C_M:1.9-2.8x10^{-4}; C_p:1.4-7.0x10^{-4}; K:0.6-0.8. The correct

TABLE 1

Commercial polyvinyl acetate standards PVAc samples distributed by Aldrich Chemical Co.

PVAc - Lot 1	PVAc - Lot 3
$\bar{M}_N = 0.0834 \times 10^6$	$\bar{M}_N = 0.103 \times 10^6$
$\bar{M}_W = 0.331 \times 10^6$	$\bar{M}_W = 0.840 \times 10^6$

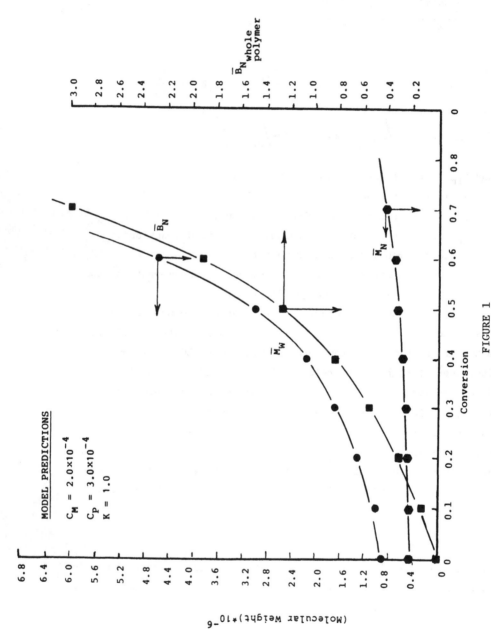

FIGURE 1

Solution of Kinetic Equations for LCB in Bulk VAc Polymerization.

kinetic parameters are expected to lie approximately within these limits.

Figures 2 and 3 illustrate the results of the MWBD methods as applied to PVAc-Lot 1 and PVAc-Lot 3 using an epsilon of 1.0. The measured $\overline{M}_N$ and $\overline{M}_W$ values are in reasonable agreement with those quoted by the manufacturer. The choice of an epsilon of 1.0 was based on the agreement of the $\overline{M}_N$ and $\overline{B}_N$ measured using the MWBD method and $\overline{M}_W$ measured by light scattering photometry $(\overline{M}_W=1.6 \times 10^6)$ for PVAc-E4. Table 3 illustrates the effect of epsilon on $\overline{B}_N$ of the whole polymer, sample PVAc-E4. $\overline{M}_N$ and $\overline{M}_N$ of the whole polymer are not affected by epsilon. An epsilon of unity appears to give adequate agreement between kinetic theory, the MWBD method and light scattering photometry. For the above, the Mark-Houwink constants used were $K = 5.1 \times 10^{-5}$ and $a = 0.791$, respectively.

TABLE 2

Solution of kinetic equations for LCB in bulk VAc polymerization.*

$C_M \times 10^4$	$C_P \times 10^4$	K	$\overline{M}_N \times 10^{-6}$	$\overline{M}_W \times 10^{-6}$	$\overline{B}_N$
2	1.5	0.6	0.54-0.58	1.7-2.2	0.6-0.9
2	2.5	0.6	0.54-0.58	2.1-2.7	0.9-1.2
2	3.0	1.0	0.63-0.71	3.0-4.6	1.3-1.9
2	4.0	1.0	0.63-0.71	3.5-5.4	1.5-2.3
2	5.0	1.0	0.63-0.71	3.9-6.2	1.8-2.8

*Values shown are for conversions of 50 and 60%, respectively.

TABLE 3

Sample PVAc-E4 - Effect of Epsilon on $\overline{B}_N$ of the whole polymer.

Epsilon	$\overline{B}_N$	$\overline{M}_N$	$\overline{M}_W$
0.5	2.5	0.51×10^6	0.137×10^7
0.8	1.4	0.51×10^6	0.137×10^7
1.0	1.1	0.51×10^6	0.137×10^7
1.2	0.34	0.51×10^6	0.137×10^7

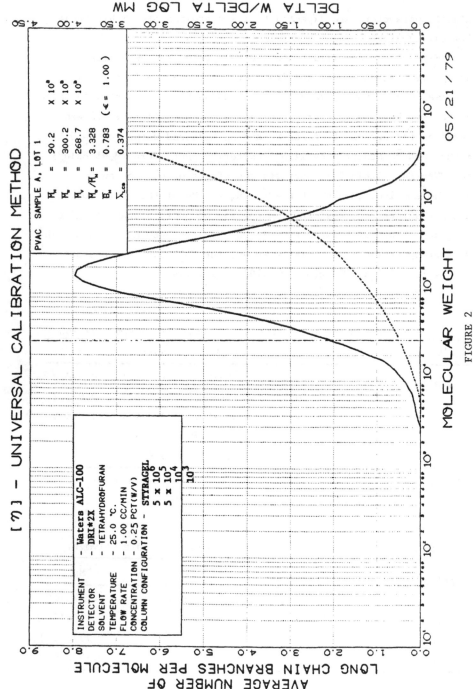

FIGURE 2

PVAc, Sample A, Lot 1.

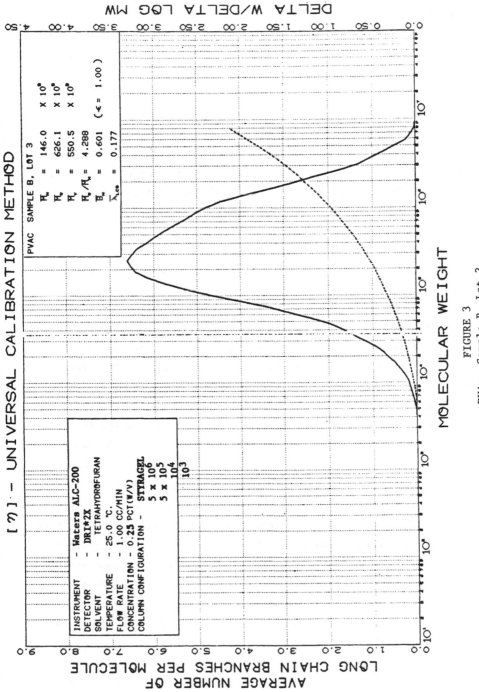

FIGURE 3
PVAc, Sample B, Lot 3.

Polyethylene Studies

Table 4 summarizes the property data for the three high pressure, tubular reactor LDPE's used to illustrate the utility of the
MWBD method. Figures 4 to 6 present the branching frequency for
these three LDPE's in terms of the number of LCB points per 1000
carbon atoms (λ_N) as a function of $\log_{10}$ MW. The number-average
number of LCB per 1000 C ($\overline{\lambda}_N$) and number-average number of LCB per
molecule ($\overline{B}_N$) values as calculated using equations (21) and (23)
are also given in Figures 4 to 6. For the MWBD method, the intrinsic viscosity data were measured at 140°C in 1,2,4-trichlorobenzene
using a Ubblelohde capillary viscometer. The Mark-Houwink constants used for polyethylene were K,a=0.51x10^{-3}, 0.706-respectively.
The chromatographic data were obtained using a modified DuPont 830
liquid chromatograph operated under the conditions shown in Figures
4 to 6.

It has been shown in a parallel study[23] with a larger group
of HP-LDPE's that an epsilon value close to 0.75 gives the best
agreement in $\overline{\lambda}_N$ values between the MWBD and C-13 NMR methods for
the whole polymer. This is illustrated in Figure 7. Therefore,
the $\overline{\lambda}_N$ values in Table IV derived using the MWBD method are in good
agreement with the values that were determined using C-13 NMR and an
epsilon value, ε=0.75. In addition, Wagner and McCrackin[24] have
reported LCB frequencies for fractions of the National Bureau of
Standards SRM #1476. Their data for these fractions are shown in
Figure 8 with our LCB frequency-MW distribution results using the
MWBD method with the #1476 standard. Excellent agreement is shown
between these two independent analyses, further validating the MWBD
method. It is also interesting to note that LDPE A was produced
at a higher conversion level than LDPE C with LDPE B immediate but
closer to the higher conversion LDPE C sample. In Figures 5 to 7,
the $\overline{\lambda}_N$ values shown are consistent with the kinetic expectations for
free radical polymerization of polyethylene to varying conversion
levels.

The shear flow behavior of the three LDPE's (A,B, and C) at
low shear rates (0.01 to 1.0 sec^{-1}) were determined using a cone

TABLE 4

Property summary for high pressure, low density polyethylenes.

Materials	Melt Index	Die Swell, %	Density, gm/cc	Intrinsic Viscosity	Blown Film Properties (38 µM thickness)			LCB Parameters		
					Haze, %	Gloss	Tear Strength, gms	$\bar{\lambda}_N$	λ_{NMR}	$\bar{B}_N$
LDPE A	1.8	72	0.919	0.796	19	53	85	4.5	4.5	5.0
B	2.6	48	0.921	0.825	6.6	122	125	2.6	2.2	2.9
C	2.1	39	0.923	0.925	4.8	156	200	1.7	1.8	2.4

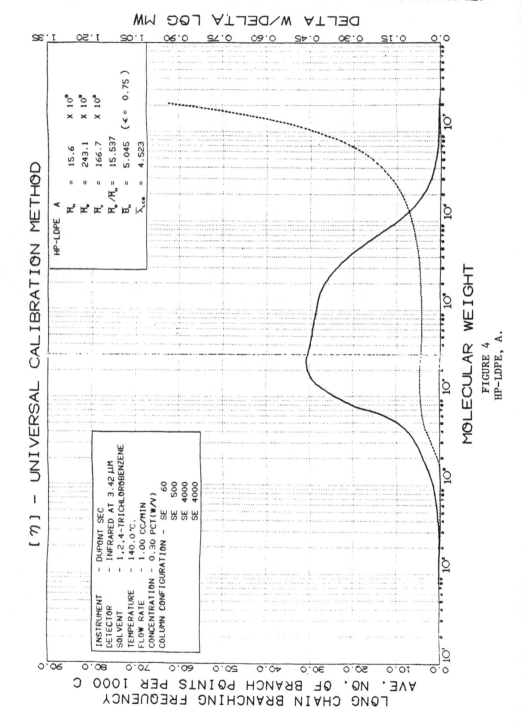

FIGURE 4
HP-LDPE, A.

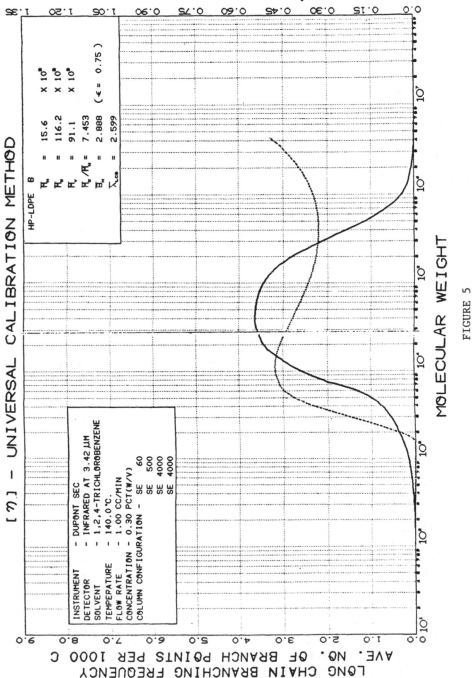

FIGURE 5
HP-LDPE, B.

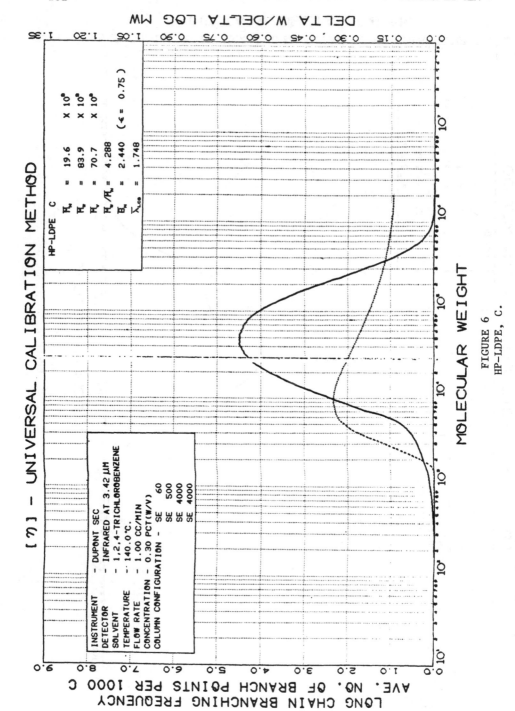

FIGURE 6
HP-LDPE, C.

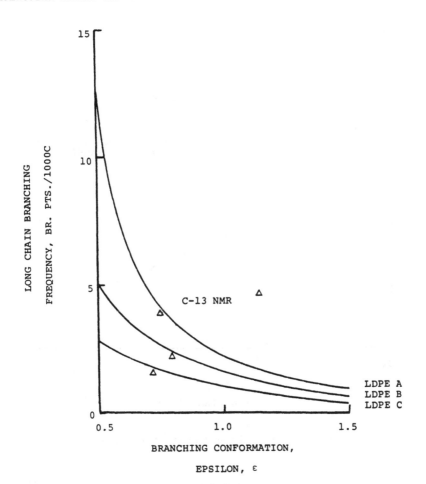

FIGURE 7
Effect of Epsilon on Long Chain Branching Frequency.   IV - Universal
Calibration Method.

and plate rheometer. The results are shown with low shear viscosity
and first normal stress difference as a function of shear rate in
Figures 9 and 10. The viscosity data approach zero shear viscosity
conditions below a shear of 0.5 sec^{-1} with the broader MW distribution LDPE A showing a lower critical shear rate for non-Newtonian
behavior. The zero shear viscosity values would not be expected to
correlate with the LCB parameters, $\bar{\lambda}_N$ and $\bar{B}_N$. Others[7] have shown
that $\log_{10}$ (zero shear viscosity) for LCB LDPE's is directly pro-

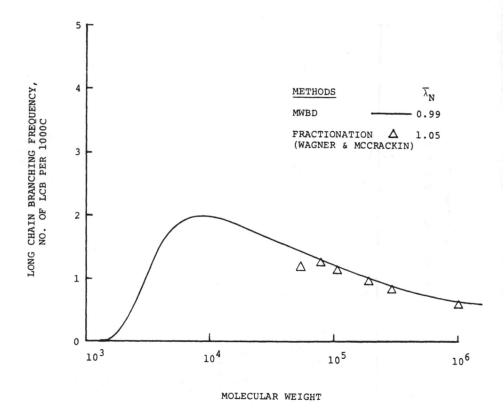

FIGURE 8

Comparison of Long Chain Branching Results for the NBS SRM #1476.

portional to the $\log_{10}$ of the product of $\overline{M}_W$ and the intrinsic viscosity ratio given by equation (18).

For thermoplastics such as PVAc or polyethylenes, chain entanglements increase with MW. These entanglements can slip and remake
under an applied stress providing a mechanism for energy storage.
Thus, melt elasticity would be expected to increase with increasing
LCB for HP LDPE's of similar low shear melt viscosities or melt
indices. This is shown in Figure 10 and by the die swell values in
Table IV.

For HP-LDPE's of similar densities (or crystallinities), the
predominant factor affecting their blown film optical properties is
surface haze or gloss due to extrusion roughness called extrusion

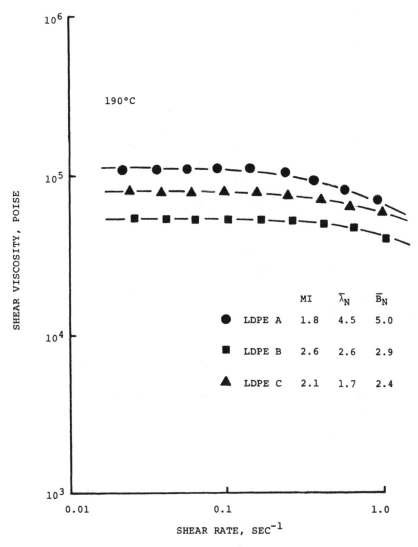

FIGURE 9
Low Shear Viscosity of LDPE's (A, B, & C).

haze.[23] Scattering-i.e., external and internal haze, due to
crystalline effects-is of minor importance. Therefore, the correlation of measured film gloss with LCB as shown in Figure 11 is to be
expected. Tear strength of blown HP LDPE films is related to the
crystalline macrostructure formed under stress or orientation condi-

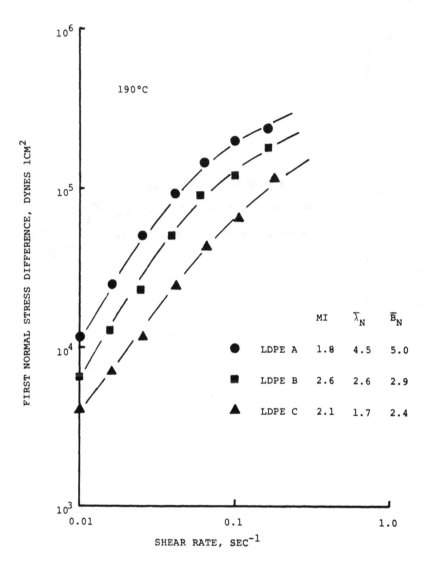

FIGURE 10
Elasticity (First Normal Stress Difference) as Effected by LCB.

tions during processing. As seen in Figure 12, increased LCB reduces
film tear strength at similar low-shear melt viscosities.

Finally, the reproducibility of the MWBD method using high-
speed, SEC and Ubbelohde viscometric data is shown in Table 5. In
general, the reproducibility of the data is within $\pm$10 percent.

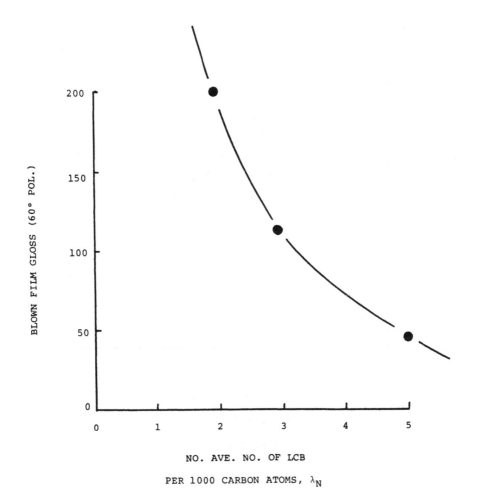

FIGURE 11
Blown Film Gloss as Effected by LCB for LDPE's of Similar Cyrstallinities.

CONCLUSIONS

1. The MWBD method provides for rapid, whole-polymer $\overline{M}_N$ values
    and for LCB-MW distribution determination. However, the
    higher MW moments - i.e., $\overline{M}_W$, $\overline{M}_Z$, and $\overline{M}_{Z+1}$ are low in value
    as a consequence of the universal calibration curve yielding
    the instantaneous $M_N(V)$ values across the chromatogram. Thus,
    the MWBD method has a significant advantage over earlier meth-

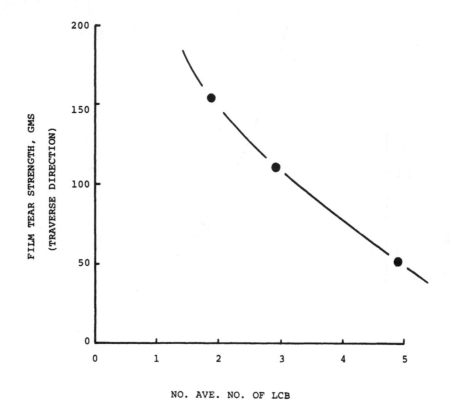

NO. AVE. NO. OF LCB

PER 1000C CARBON ATOMS, λ_N

FIGURE 12
Blown Film Tear Strength and LCB Inter-Dependence.

ods proposed by Drott[7] and Ram and Miltz[8] for LCB information.

2. The MWBD method requires an absolute measure of epsilon. C-13
    NMR spectroscopy can provide this data by fitting experimental
    LCB parameters from the MWBD method - i.e., assuming different
    epsilon values, calculating the $\overline{\lambda}_N$ value using the MWBD method,
    and comparing these calculated branching frequencies to those
    from C-13 NMR.

3. The MWBD method is sensitive to the Mark-Houwink constants for
    the linear homologue used in the analysis. Reliable values
    are essential to establish realistic branching frequencies.

TABLE 5

NBS SRM 1475 and 1476 results.

Materials		Molecular Weight Parameters			LCB Frequency
		$\overline{M}_N$ x10^{-3}	$\overline{M}_W$ x10^{-3}	$\overline{M}_W/\overline{M}_N$	$\overline{\lambda}_N$
NBS SRM #1476		22.44	86.34	3.85	1.42
		23.70	91.73	3.87	1.55
		21.30	86.47	4.06	1.39
		24.57	88.36	3.60	1.50
	Avg.	22.99	88.23	3.83	1.47
NBS SRM #1475		17.93	52.58	2.93	0.0
		17.94	53.45	2.97	0.0
	Avg.	17.94	53.02	2.96	0.0

4. For polyvinyl acetates, the use of $\varepsilon = 1.0$ with the MWBD method gives good agreement for the whole polymer branching frequencies with the kinetic theory. Further kinetic experiments are required to optimize epsilon.

5. The LCB-MW distribution results obtained with the MWBD method for bulk synthesized PVAc's are as would be expected based on free radical kinetic considerations. Further kinetic experiments are required to optimize kinetic parameters, C_M, C_p and K.

6. For polyethylenes, the use of $\varepsilon=0.75$ with the MWBD method gives excellent agreement for the branching frequencies when compared with C-13 NMR results. Earlier works by Drott[7] and Ram and Miltz[8] used epsilon values of 0.5. In an extensive literature review, Small[25] found general agreement that epsilon should have a value of 1 ± 0.3. Epsilon values near 0.75 are reasonable for the type of LCB in high pressure, free radical polymerized, LDPE's-i.e., branching of random lengths and spacings between branch points.

7. The LCB-MW distribution results obtained with the MWBD method for typical commercial, high pressure process LDPE's are as would be expected based on free radical kinetic considerations.

8. The LCB-MW distribution for the National Bureau of Standards SRM #1476 is in good agreement with fractionation data obtaied by McCrackin and Wagner.[24] Reproducibility studies using the MWBD method with the NBS SRM #1476 gave $\overline{\lambda}_N$ and $\overline{B}_N$ values to ± 10 percent and M_N's within ± 5 percent using high-speed, SEC techniques.

9. Both LCB parameters ($\overline{\lambda}_N$ and $\overline{B}_N$) for high pressure LDPE's show excellent correlation to melt elastic properties and to blown film properties such as opticals and tear strength.

10. The MWBD method, when coupled with high-speed SEC techniques, is more rapid for LCB frequency measurement, has a greater sensitivity at low LCB levels and has improved reproducibility than can be obtained with present C-13 NMR instrumentation. In addition, LCB-MW distribution information is provided which C-13 NMR cannot do without laborious polymer fractionation.

ACKNOWLEDGEMENTS

Dr. E. S. Hsi (Union Carbide Corporation, Bound Brook, N. J.) provided the C-13 NMR results on the three HP-LDPE's studied.

REFERENCES

1. Roedel, M. J., J. Amer. Chem. Soc., 75, 6110 (1953).

2. Wild, L., Ranganath, R. and Knobeloch, D. C., Polymer Eng. and Sci., 16, 811 (1976).

3. White, J. L., "Experimental Study of Elongation Flow of Polymer Melts," Polymer Science and Engineering Report, No. 104, University of Tennessee, Knoxville, Tennessee, July, 1977.

4. Mirabella, F. M. and Johnson, J. F., Macromol, J., Sci-Revs. Macromol. Chem., 612(1), 81-108 (1975).

5. Grubisic, Z., Remp, P., and Benoit, H., J. Polym. Sci. B, 5, 753 (1967).

6. Hamielec, A. E. and Quano, A. C., J. Liq. Chromat., 1, 111 (1978).

7. Drott, E. E. and Mendelson, R. A., Sci. A-2, 8, 1373 (1970).

8. Ram, A. and Miltz, J., J. Appl. Polym Sci., 15, 2639 (1971).

9. Williams, G. R. and Cervenka, A., Eur. Polym. J., 8, 1009 (1972).

10. Wild, L. and Guliana, R., J. Polym. Sci. A-2, 5, 1087 (1967).

11. Meyerhoff, G., Separation Sci., 6, 239 (1971); Goedhart and Opschoor, A., J. Polym. Sci A-2, 8, 1227 (1970).

12. Ouano, A. C., J. Polym. Sci A-1, 10, 2169 (1972).

13. Barlow, H., Wild, L., and Ranganath, J. Appl. Polym. Sci., 21, 3319 (1977); ibid., 21, 3331 (1977).

14. Ouano, A. C. and Kaye, W., J. Polym. Sci. A-1, 12, 1151 (1974).

15. Ouano, A. C.,

16. MacRury, T. B. and McConnell, M. L., J. Appl. Polym. Sci., in press (1979).

17. Zimm, B. H. and Kilb, R. W., J. Polym. Sci., 37, 19 (1959).

18. Morton, M., Helmmiak, T. E., Gadrary, S. D. and Buiche, F., J. Polym. Sci., 57, 471 (1962).

19. Noda, I., Horikawa, T., Kato, T., Fryimoto, T. and Nagasawa, N., Macromolecules, 3, 795 (1970).

20. Graessley, W. W., "Detection and Measurement of Branching in Polymers," Characterization of Macromolecular Structure, Mc-Intyre, D., Ed., Natl. Acad. of Sci. Publication No. 1573, Washington, D. C., pp. 371-388 (1968).

21. Zimm, B. H. and Stockmayer, W. H., J. Chem. Phys., 17, 1301 (1949).

22. Graessley, W. W., Uy, W. C. and Gandhi, A., I & EC Fundamentals, 8, 696 (1969).

23. Foster, G. N., "Characterization of Long Chain Branching in Polyethylene," presented at the 13th ACS Middle Atlantic Regional Meeting, Monmouth College, West Long Branch, N. J., March 1979.

24. Wagner, H. L., and McCrackin, F. L., J. Appl. Polymer Sci., 21, 2833 (1977).

25. Small, P. A., Advances in Polymer Science, Fortschritte der Hochpolymeren-Forsclang, Vol. 18, Springer-Verlag, New York (1975).

GEL PERMEATION CHROMATOGRAPHY OF NITROCELLULOSE

A.F. Cunningham, C. Heathcote, D.E. Hillman, J.I. Paul

Materials Quality Assurance Directorate
Royal Arsenal East
Woolwich, UK

ABSTRACT

This paper reviews problems encountered in the gel permeation chromatography of propellant grade nitrocellulose (12.6% N), blasting explosive grade (12.2% N) and cellulose nitrated by a mild non-degrading method (13.4% N). Quantitative comparisons show the effect of concentration, calibration method, choice of column exclusion limits and the degree of nitration. The use of low angle laser light scattering to overcome problems in calibration and to show the effect of microgel on the separation is also described.

The changes in molecular weight during the stages of the industrial nitration process are described.

INTRODUCTION

The majority of the many papers published in the open literature on the gel permeation chromatography (GPC) of nitrocellulose arise from studies of the molecular weight distribution of cellulose. Because of the insolubility of cellulose in the normal GPC solvents it has been necessary to prepare soluble derivatives, e.g., acetate, carbanilate or, most commonly, nitrate. These are soluble in solvents such as tetrahydrofuran and, provided that the process of derivative formation does not cause degradation of the cellulose, the resultant GPC chromatograms can be regarded as representative of the cellulose itself.

Laboratory nitration under mild conditions[1] gives a high nitrogen content (13.4% N) which is close to full nitration (14.1% N). Industrial grades of nitrocellulose used in propellant manufacture and in blasting explosives have a lower nitrogen content and have considerable residual hydroxyl content which might affect the GPC behaviour.

This programme of work began as an investigation into problems in processing of double base propellant, which coincided with a change in supplier of the cotton linters used in the manufacture of the cellulose nitrate component of the propellant. Comparison of batches of nitrocellulose showed that GPC was the only analytical technique to show differences between batches giving good and bad processability. Therefore, molecular weight distributions were determined for a wide range of samples, including samples from individual stages of manufacture, from studies of the effect on the propellant of systematic changes in the manufacturing process, and from different suppliers of cotton linters.

Study of the published GPC work on nitrocellulose has shown some disagreement between authors and a lack of information in certain aspects of the technique which have appeared to be important in our work. This report discusses these points and highlights problems which may be encountered.

Segal[2] has published a very comprehensive review on the GPC of cellulose and derivatives. No attempt is made in this paper to give a full range of references; those cited are considered to be most appropriate to the discussions.

EXPERIMENTAL

Gel Permeation Chromatography

Unless otherwise stated, a Waters Associates Model 200 GPC unit was used with 4 ft x 3/8" 'Styragel®' columns and refractometer detector. Two main sets of operating conditions were used:

	Series I	Series II
Column exclusion limits	$10^6 + 3 \times 10^5 + 3 \times 10^4 + 10^3$	$5 \times 10^6 + 10^5 + 3 \times 10^4 + 3 \times 10^3$
Solvent	Tetrahydrofuran	Tetrahydrofuran
Flow Rate	1 ml/min	1 ml/min
Solution Strength	1/8% and 1/4% using air dried nitrocel-lulose	1/8% using air dried or vacuum dried nitro-cellulose
Sample solution preparation	Allowed to stand over-night, filtered through Kruger pad	Allowed to stand over-night, filtered through Kruger pad + Whatman GF/F filter paper (0.7μ)
Sample size	2 ml	2 ml
Refractometer span	X8	X16
Syphon size	5 ml	2.5 ml
Expression of results	Expressed in terms of Å derived from poly-styrene calibration with extrapolation of molecular weight aver-ages to zero concentra-tion.	As molecular weight de-rived from polystyrene calibration and use of Q- factor of 10.5

Where an ultra-violet detector (254 nm) was used, an Applied Chromatography Services instrument was placed in series with the normal
refractometer detector.

Light Scattering (LALLS)

All work was carried out using a Chromatix KMX-6 low angle
laser light scattering photometer in either the static mode or with
a flow-through cell in-line with the GPC. The KMX-6 cell was then
placed between the column outlet and the refractometer cell.

The detailed method of use has been described.[4,5] When used
in-line with the GPC, the outputs of both GPC and LALLS were recorded.
At a given elution volume, the refractometer response was used to
calculate the polymer concentration in the LALLS cell. The excess
Rayleigh scatter, recorded from the LALLS, was corrected to zero
concentration using the second virial co-efficient of the equation

relating Reyleigh scatter to concentration. The polymer molecular
weight was ten calculated using the refractive index increments of
Badger and Blaker[3].

RESULTS AND DISCUSSION

The report is divided into two sections:

Section 1   Choice of Instrumental parameters and calibration
             methods

Section 2   Study of the industrial process for production of
             mechanical pyro grade nitrocellulose (12.6% N)

1a. Choice of Column System

Although native cellulose is claimed to be monodisperse, the
cotton linters used for nitrocellulose manufacture are chemically
treated, e. g., by heating with sodium hydroxide and by bleaching.
These treatments give the required purity and viscosity but lead to
broadening of the molecular weight distribution. Further changes
occur during the nitration process, leading to the broad chromatograms shown in Figures 1 and 2 (GPC conditions Series II). Series
II, which included a nominal 10^7 Å Styragel column, showed unexpec-
tedly better separation than Series I in the high molecular weight
region, with signs of a long leading edge to the envelope. Some
materials, notably the celluloid grade (11.0% N) and the blasting
explosive grade (12.2% N), required addition of low exclusion limit
columns to separate the low molecular weight nitrocellulose from the
solvent impurity peak ($10^6 + 3 \times 10^5 + 3 \times 10^4 + 10^3 + 500 + 100$ Å
used for blasting explosive). Very high detector stability was required at maximum sensitivity to define clearly the beginning and
end of the nitrocellulose chromatogram. Values of $\overline{M}_n$, in particular,
are sensitive to small uncertainties in base line position and are
sometimes suspect.

Microparticulate columns of high exclusion limit (10^7 Å) were
not available until recently, and therefore have not been normally
used in this programme.

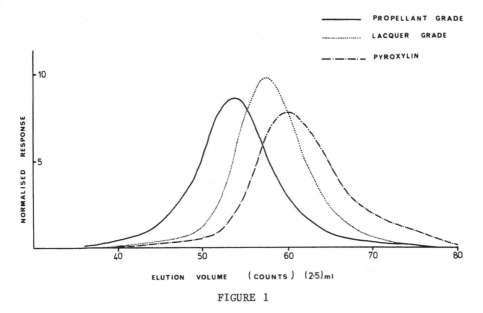

FIGURE 1

Typical Nitrocellulose Chromatograms.

1b. Calibration Standards

The conventional calibration, by use of narrow range polystyrene standards, followed by conversion to molecular weight by means of one of the universal calibration methods, is widely used for nitrocellulose. As shown in Figure 2 however, the normal polystyrene standards of molecular weight up to 3×10^6 require excessive extrapolation of the calibration to cover the highest molecular weight components of most samples of industrial nitrocellulose, although the degree of polymerisation (DP) of these samples is not unusually high.

1c. Concentration Effect

The shift of elution volume of a polymer with change in concentration is well recognised. With nitrocellulose, the effect is very large even at low concentration. In early work, solutions were run at 1/4% and 1/8% concentrations and results extrapolated to zero concentration. This procedure was excessively tedious and, for

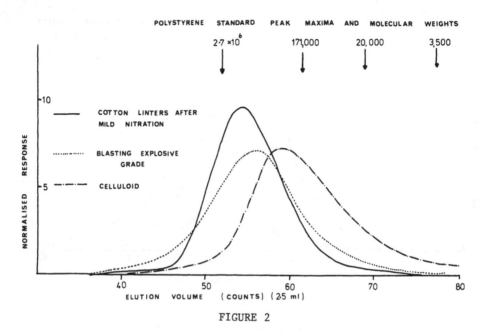

FIGURE 2

Typical Nitrocellulose Chromatograms.

Series II conditions, a single determination at 1/8% concentration
was used. This is the most usual procedure of other workers.

Figure 3 shows a typical plot of $\overline{A}w$ against concentration.
For propellant grade material (12.6% N) the plot is nearly linear
and the effect of concentration is less important. For cotton linters, nitrated by mild laboratory conditions, the effect is much
greater. Assuming the linear extrapolation to be valid, failure to
correct to zero concentration would lead to errors in weight average
results ($\overline{A}w$) as below:

	Lab nitrated linters (13.4% N)	Propellant grade (12.6% N)
1/8% Concentration	27,000	9,700
Zero Concentration	36,000	10,400
% error at 1/8% Concentration	23	7

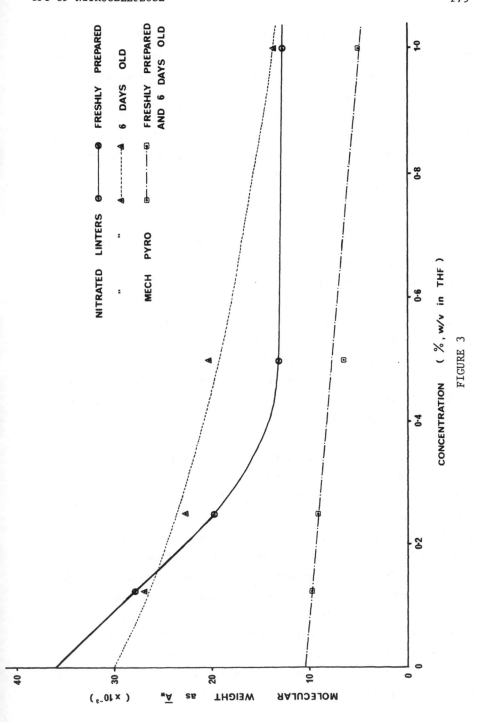

FIGURE 3

Nitrocellulose – Concentration and Aging Effects.

Whilst the assumption of a linear relationship between $\overline{A}w$ and concentration may not be fully correct, and the extrapolation based on two concentrations only may be criticised, the procedure appears to minimise errors due to concentration effects.

The concentration effect increases greatly with molecular weight. Many workers have investigated nitrocellulose at much higher degree of polymerization than the DP_w of approx. 500-1200 used here. The effect of concentration on studies of molecular weight calculation methods may clearly be critical.

1d. Stability of Solutions

The work of Segal[7] has shown that DP_w for nitrocellulose in tetrahydrofuran showed no change after 10 days and little change over 54 days. Carignen and Turngren[8] showed that light scattering results in acetone remained unchanged for 14 days, although ethyl acetate solutions degraded from $\overline{M}_w$ = 200,000 to 140,000 in the same period. Propellant grade nitrocellulose showed negligible change in $\overline{A}_w$ at 1/8% concentration in tetrahydrofuran. A decrease from 38 x 10^3 to 30 x 10^3 in 6 days, for nitrated linters, was probably partly attributable to the concentration effect, the values at 1/8% being closely similar (Fig. 3).

1e. Presence of Microgel

The use of the higher exclusion limit column system has shown small but significant amounts of very high molecular weight nitrocellulose which are eluted very early. This effect appears to be repeatable and not influenced greatly by changes in filtration procedure. Light scattering - GPC shows the presence of microgel in acetone (Fig. 4) which gives a strong response at about the exclusion limit. The lack of refractometer response suggests that the GPC refractometer response is that of polymer in true solution. Similar work in tetrahydrofuran led to blockages of filters before the light scattering cell although the sample later passed through the refractometer. Analyses in acetone showed that the exclusion limit peak on the light scattering response could be reduced by filtration through finer filters. Static measurements of weight-

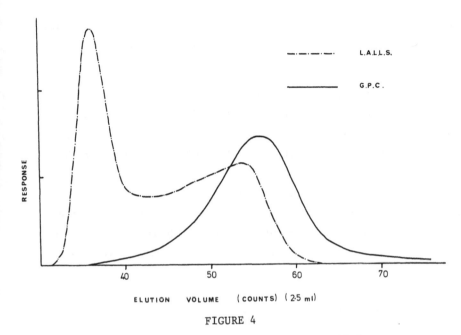

FIGURE 4

L.A.L.L.S./G.P.C. Analysis of Nitrocellulose.

average molecular weight by light scattering have given great difficulty due to the presence of microgel, indicated by non-repeatable
response and very high molecular weight. Comparison of molecular
weight by GPC therefore appears to offer a big advantage over light
scattering in that microgel presence has relatively little effect.

1f. Light Scattering for Calibration of GPC Column in Acetone

By slowly conditioning GPC columns from tetrahydrofuran to
acetone, using stepwise changes at slow flow rate (i. e. 100:0,
75:25, 50:50, 25:50 and 0:100% blends) the LALLS instrument was
used to determine an actual molecular weight calibration curve for
12.2% and 12.6% N grades. Direct comparison with a polystyrene
calibration was not possible due to the insolubility of polystyrene.
However, by comparing a cumulative distribution for a propellant
grade sample in both acetone and tetrahydrofuran, a relationship
was obtained for LALLS molecular weight in acetone against elution
volume and polystyrene extended chain length in tetrahydrofuran. An

experimental Q-factor was obtained from this plot at high molecular
weights and showed rough agreement with that of Chang[4]. This
point is further discussed below.

1g. Influence of Degree of Nitration

Segal[9] investigated the effect of degree of nitration on
molecular weight distribution data for the range 13.51 - 13.81% and
concluded that there was little effect on GPC behaviour over this
limited range.

The work of Lindsley and Frank[10] give an equation relating
intrinsic viscosity for cellulose trinitrate in acetone with that
for other lower nitrogen contents. Figure 5 shows the calibrated
value of ηM (intrinsic viscosity x molecular weight) against nitro-

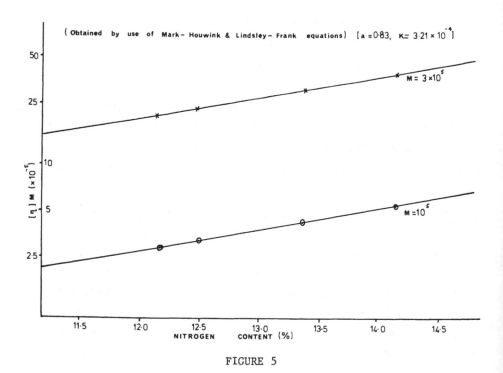

FIGURE 5

Nitrocellulose - Relationship Between Percentage Nitrogen Content
and Log $[\eta]M$. (Using Mark-Houwink and Lindsley-Frank Equations;
$a = 0.83$, $k = 3.21 \times 10^4$).

gen content, assuming that the same equation is valid in THF solutions. Since elution volume in GPC is defined by ηM, large errors should be introduced by failing to take account of the variation in nitrogen content. Thus, for typical nitrocellulose grades:

% Nitrogen:	14.15	$\eta M = 40 \times 10^{-5}$, approximately
	13.4	32 " "
	13.1	28.5 " "
	12.6	24.5 " "
	12.2	21.7 " "
	11.5	17.3 " "

There appears to be no corresponding information in the literature on the variation of intrinsic viscosity and nitrogen content in THF solution. However French[11] used this relationship assuming similar behaviour in THF and acetone.

Attempts to briefly study this effect in THF by mild nitration[1] of a cotton sample to give nitrogen contents of 13.3% and 12.6% showed negligible difference in peak elution volumes:

| 13.3% nitrogen | Calculated peak mol wt: | 189×10^3 |
| 12.55% " | " " " " : | 168×10^3 |

For the same degree of polymerisation, a molecular weight difference of 7×10^3 would arise from the additional nitrate content, i.e., the effect on the GPC separation is only 14×10^3 or 8%. Complete solubility in THF appears to require approximately 11% or more nitrogen, i.e., a degree of substitution corresponding to the dinitrate.

Comparison of refractometer and ultra-violet detector responses ιor propellant grade (12.6% N) samples have shown identical normalised chromatograms although refractive index would be related to weight of polymer and ultraviolet response to nitrate content. The samples therefore appear to be homogeneous in terms of nitrogen content. Evidence in this laboratory therefore suggests, tentatively, that no major changes in elution volume with nitrogen content take place over the range 12.55 - 13.3% N.

1h. Calibration for Molecular Weight

Characterisation of nitrocellulose by GPC has been carried out by a number of workers in the past and considerable difficulty has been encountered in establishing valid molecular weight figures. Ideally, GPC calibration should be carried out with well-characterised narrow distribution fractions of the polymer under test. Where such fractions are not available, the instrument is calibrated with polystyrene standards and conversion from polystyrene molecular weights to those of the test polymer is made by one of several 'universal' calibration methods. These are based on the use of extended chain length, hydrodynamic volume or unperturbed dimensions of the polymer and of polystyrene. A batch of propellant grade nitrocellulose was examined by GPC using Series II operating conditions and molecular weight calculated by the 'universal' calibration methods listed below:

Q-factor method

Segal[6], in early work, calculated a Q-factor of 58 for cellulose trinitrate, based on X-ray measurements. Lower values were calibrated for lower degrees of nitration, e. g., $Q = 55.9$ at 13.5% N. Normally, this appraoch is used only when there is a similar degree of interaction between the GPC solvent and polystyrene as between the solvent and polymer under test. This is clearly not applicable in the case of nitrocellulose.

Effective density factor ($\overline{Q}$)

Chang[4] has suggested the use of this modified Q-factor approach. $\overline{Q}$ is obtained by dividing the number average molecular weight ($\overline{M}_n$), obtained by osmometry, by the peak extended chain length, obtained by GPC of polystyrene standards. By using a range of nitrocellulose samples a factor of $\overline{Q} = 10.5$ was obtained. This factor is therefore based on properties in solution rather than in the solid state.

This approach gave much lower molecular weights than the Q-factor method and the results were in broad agreement with those obtained by other methods (Table 1). However, the assumption that peak extended chain length corresponds with $\overline{M}_n$ is an approximation which may invalidate the precision of this otherwise logical approach.

Hydrodynamic volume

The 'universal' calibration, based on hydrodynamic volume, was originally proposed by Benoit[12]. A plot of elution volume against log $(\eta)M$ gives a calibration which is valid for a very wide range of polymers. $(\eta)M$ is called the 'coil size', which is a measure of the hydrodynamic volume, (η) is the intrinsic viscosity and M the molecular weight. Polystyrene is used to calibrate the GPC in terms of $(\eta)M$ and a number of approaches may be used to evaluate the nitrocellulose molecular weight.

By combining the Benoit hydrodynamic volume relationship with the Mark Houwink equation, the following formula is obtained for a given elution volume:

$$\text{Log } M_p = \frac{1}{1 + a_p} \log \left[\frac{K_{ps}}{K_p} \right] + \frac{(1 + a_{ps})}{(1 + a_p)} \text{ Log } M_{ps}$$

Where   M = molecular weight

        a, k = Mark-Houwink constants

      PS, P = Subscripts corresponding to polystyrene and polymer under test respectively

Therefore, if values of 'a' and 'K' are known, a polystyrene calibration can easily be transposed for use with another polymer. However, the values of these constants quoted in the literature vary very greatly and are said to be dependent on the degree of polymerisation.

Molecular weight averages for the nitrocellulose sample are given in Table 1 using different literature values and 'a' and 'K'. In general, they show reasonable agreement, although the values of Jenkins[13] are clearly low whilst those of French[11] are high. The relatively low values of D_p with our samples may favour consistent results by these different calibration methods since this avoids the

TABLE 1

Molecular Weights for Propellant Grade Sample Calibrated by Different Calibration Methods. GPC Operating Conditions - II

	Method	Data	$\overline{M}_w$ x 10^{-4}	$\overline{M}_n$ x 10^{-4}	$\dfrac{\overline{M}_w}{\overline{M}_n}$	Reference
1.	Q-factor	Q = 54.1	127.4	24.7	5.1	6
2.	Q-factor of CHANG	$\overline{Q}$ = 10.5	24.7	4.8	5.1	4
3.	Hydrodynamic volume	'a' & 'K' values of SEGAL with	a = 0.76, K = 82.6x10^{-5}) N/C D_p >1000 ; a = 1.14, K = 0.7x10^{-5}) N/C D_p <1000 ; Polystyrene value as a (4)				13
4.	"	'a' & 'K' values of SEGAL. Zero shear. D_p 60-6000	a = 0.83, K = 32.1x10^{-5}) N/C ; a = 0.71, K = 13.4x10^{-5}) Polystyrene	22.7	5.5	4.1	13
5.	"	'a' & 'K' values of OUANO for polystyrene, JENKINS for N/C	a = 0.65, K = 28.9x10^{-5}) Polystyrene ; a = 1.014, k = 6.06x10^{-5}) N/C	14.7	5.1	2.9	14 ; 12

	Method	Data	$\bar{M}_w \times 10^{-4}$	$\bar{M}_n \times 10^{-4}$	$\dfrac{\bar{M}_w}{\bar{M}_n}$	Reference
6.	Hydrodynamic volume	$\left.\begin{array}{l} a = 0.89 \\ K = 21.9 \times 10^{-5} \end{array}\right\}$ N/C	19.1	5.0	3.8	4
	'a' & 'K' values of CHANG	$\left.\begin{array}{l} a = 0.72 \\ K = 12.6 \times 10^{-5} \end{array}\right\}$ Polystyrene				
7.	"	$\left.\begin{array}{l} a = 0.84 \\ K = 18. \times 10^{-5} \end{array}\right\}$ N/C	29.5	7.1	4.2	10
	'a' & 'K' values of FRENCH. (N/C) and SPATORICO for polystyrene	$\left.\begin{array}{l} a = 0.725 \\ K = 11.1 \times 10^{-5} \end{array}\right\}$ Polystyrene				15
8.	Light Scattering	In line with GPC in acetone	24.8	–		
			22.9	–		
9.	"	Static measurement	29.3	–		

alleged discontinuity at about D_p = 1000 in the relationship for 'a' and 'K' against D_p. Table 1 shows the results obtained by calibration of molecular weight averages for a single batch of propellant grade nitrocellulose using the methods described above. Light scattering results are also quoted.

Although methods 3-7 are all based on the hydrodynamic volume calibration, they show considerable variation depending upon the literature values used for the 'a' and 'K' constants for polystyrene and nitrocellulose. The results by methods 2, 3, 4 and 6 are in broad agreement with each other, with the $\overline{Q}$-factor of Chang and with light scattering in acetone when used in-line with the GPC. The higher static light scattering result may be due to traces of microgel increasing the degree of scatter.

The use of a $\overline{Q}$-factor of 10.5 appears to be the simplest method of obtaining actual molecular weight for propellant grade nitro-cellulose (12.6% N).

2. Industrial Nitration Process

Nitrocellulose is normally prepared by the mechanical process which is outlined below:

1. Cotton linters are shredded and dried on conveyor belt at 120° F.

2. Nitration by mechanical agitation in mixed nitric-sulphuric acids.

3. Centrifuging to remove spent acids and discharged into large excess of water.

4. Stabilisation by (a) boiling with acid
                         (b) boiling with alkali
                         (c) boiling with water

5. Pulping. To bruise fibres and release impurities and to shorten fibre length.

6. Grit extraction, by centrifuging.

7. Washing and wet-blending.

8. Dewatering by centrifuging to a water content of about 30%.

Some nitrocellulose is also produced by the displacement process which is essentially similar except for stages (2) and (3), where the nitration is carried out under static conditions and the acid is removed by running a stream of water into the vat.

The exact conditions of nitration govern the final nitrogen content of the product:

13.35% nitrogen	Guncotton
12.45% - 12.65%	Pyrocellulose ('Pyro')
12.2 %	Propulsive soluble

Intermediate grades may be prepared by blending.

RESULTS AND DISCUSSION

1. Mechanical Nitration Process

This project began with the examination of samples of nitro-cellulose which gave satisfactory and unstaisfactory casting powder, followed by examination of samples taken at different stages of the manufacturing process. Correlation of results was difficult, as samples were obtained over a long period of time. Finally, a single batch of cotton linters was taken through the whole mech pyro manufacturing process with samples being taken at each stage. For convenience, this series of tests is discussed first.

The results of these tests are listed in Table 2. As described in Section 1, molecular size is expressed in terms of extended chain length (ECL) in angstrom units. Molecular weight is a controversial subject but, for the purpose of this report where comparison of chromatograms is more important than actual molecular weight, the use of ECL is simpler and satisfactory. The relationship between ECL and molecular weight for the products of industrial nitration (12.6% nitrogen) and mild laboratory nitration (13.5%) has been assumed to be similar.

All average values were obtained by running samples at two concentrations and extrapolating to zero concentration.

Cotton linters from one batch were nitrated to obtain pyro nitrocellulose. Two series of samples were taken from this manufacturing
process (A and B in Table 2).

The difference in $\bar{A}_w$ values for the cotton starting material is
greater than expected for experimental error and suggests some lack
of homogeneity. The enormous drop in $\bar{A}_w$ value on nitration (samples
4A and B) is not expected, but the very large decrease on boiling
(Samples 7A and B) is larger than might be anticipated. The remainder
of the precessing appears to give little change, the increase of molecular weight for sample 12A being probably due to non-homogeneity.
The changes in molecular weight distribution are illustrated in
Figure 6.

Series B appears to have undergone more drastic degradation than
Series A and suggests that, under normal industrial conditions, there
may be appreciable differences in molecular weight distribution or

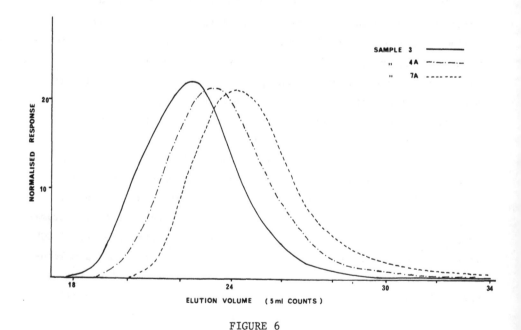

FIGURE 6

Nitrocellulose - Effect of Manufacturing Process on Molecular
Weight.

TABLE 2

Study of the Manufacturing Process (Mechanical Nitration).

Sample	Manufacturing Stage	$\bar{A}_w \times 10^{-3}$	
		A Series	B Series
1 & 2	Cotton linters (two samples).....	26.7* 33.6*	–
3	Sample from cotton bin, i.e.,.... after drying	36.4	–
4.	'Green' N/C after nitration......	20.0	17.5
7.	N/C after acid boil and first.... soda boil	8.1	8.8
9.	N/C after second soda boil.......	9.9	8.1
11.	During discharge of barrel.......	10.7	9.2
12.	N/C after pulping but before..... washing	12.3	8.1
13.	Sample of final blended N/C (blend of A and B series)	10.1	–

*Nitrated by mild laboratory procedure

average molecular weight between different nitration batches although the final blending stages will tend to reduce this effect.

2. Comparison of Cotton Linters and Methods of Nitration

Initially, differences in the cotton linter starting material were suspected as being the cause of problems in the manufacture of propellant. Therefore, cotton linters from three different sources of supply were nitrated by the mild procedure. In addition, three separate batches of cotton from one supplier were treated in a similar way. The results are shown in Table 3.

The effect of different methods of laboratorynitration and the normal industrial nitration on a single batch of cotton linters is shown in Table 4.

The mild nitration process is claimed to give no degradation. The use of nitric-sulphuric acid mixtures, however, clearly give a very high degree of degradation which is greater in the industrial process than in the equivalent laboratory method. The differences

TABLE 3

Different Sources of Cotton.

Supplier	Nitration Method	$A_w \times 10^{-3}$	$A_n \times 10^{-3}$	$\dfrac{\overline{A_w}}{\overline{A_n}}$
Firm A	Phosphoric Acid/P_2O_5/HNO_3	42.1	16.4	2.6
Firm B	" " " "	34.2	8.5	4.0
Firm C	" " " "	43.0	13.7	3.2

TABLE 4

Effect of Nitration Conditions.

Nitration Method		$\overline{A_w} \times 10^{-3}$	$\overline{A_n} \times 10^{-3}$
Phosophoric acid/P_2O_5/HNO_3,	Laboratory	42.1	16.4
Nitric acid/H_2SO_4/H_2O,	Laboratory	16.6	2.2
" " " " ,	Laboratory	9.6	1.5

in $\overline{A_w}$ between the batches of linters is small compared with the changes brought about by industrial nitration and it is, therefore, possible that the molecular weight of the linters will not appreciably influence the final properties of the propellant.

3. Effect of Boiling Period

As shown in Table 2, the stabilisation stages lead to a very large change in $\overline{A_w}$. Nitrocellulose from a batch which gave an unsatisfactory product was boiled for additional periods with the results shown in Table 5. There is a progressive decrease in $\overline{A_w}$ with time. This corresponded with a general shift of the whole GPC envelope to lower molecular weights.

TABLE 5

Effect of Boiling on Molecular Weights.

Boiling Time	$\overline{A}_w \times 10^{-3}$
Normal	13.9
Normal + 39 hours	13.2
Normal + 87 hours	10.6

4. Influence of Cotton Fibre Lengths

The results in Table 6 and Figure 7 show the influence of fibre length. Two samples from supplier A were separated into three mesh sizes and nitrated by the mild method. It can be seen that generally the values of $\overline{A}_w$ are similar but $\overline{A}_n$ decreases significantly with decrease in fibre length. Although $\overline{A}_n$ is difficult to measure by GPC with any precision, there are nevertheless considerable changes in dispersity. More low molecular weight material is clearly present in the finer fractions.

TABLE 6

Effect of Fibre Lengths on Molecular Weight.

Sample	Retained on sieve	$\overline{A}_w \times 10^{-3}$	$\overline{A}_n \times 10^{-3}$	$\dfrac{\overline{A}_w}{\overline{A}_n}$
21	BS 50	43.1	16.7	2.6
"	BS 100	46.2	10.4	4.5
"	BS 200	38.1	4.4	8.7
22	BS 50	40.7	16.2	2.5
"	BS 100	37.7	7.2	5.2
"	BS 200	41.0	9.0	4.5

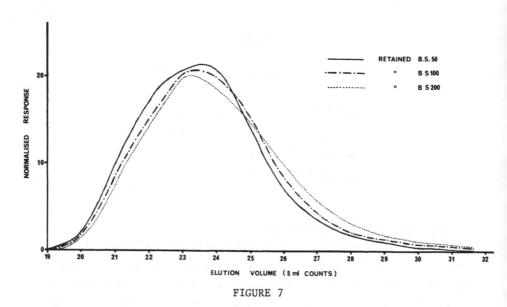

FIGURE 7

Nitrocellulose – Effect of Fibre Length on Molecular Weight Distribution.

5. Influence of Beating

Samples of one batch of nitrocellulose which had been processed by three separate beaters were examined but showed no significant differences.

6. Correlation of $\overline{A}_w$ with Viscosity Measurements

A range of batches of mech pyro were examined by GPC and the resultant values of $\overline{A}_w$ plotted against the viscosities obtained by falling sphere method (ICI reference Jan-244). See Figure 8. There appears to be reasonable correlation.

CONCLUSIONS

GPC has been shown to be a valuable tool for the study of nitrocellulose and cotton linters in industrial processing. However, the derivation of actual molecular weight distributions, as opposed to

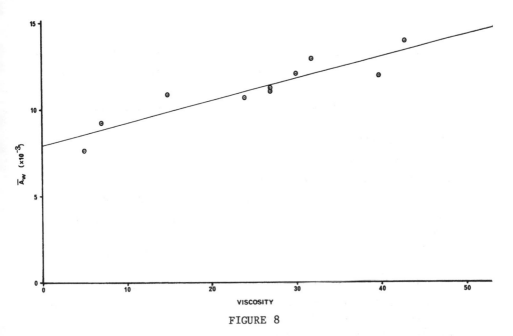

FIGURE 8

Nitrocellulose - Comparison Between Viscosity and Molecular Weight.

comparative distributions, is subject to a number of errors in technique and interpretation. Extreme care must therefore be taken in
work of this type and a great deal of work is required in the future
to allow clear cut interpretation of data. Provided these problems
are understood, the technique of GPC can be applied to industrial
problems with great success.

ACKNOWLEDGEMENTS

    Many people and organisations have contributed to this project
notably Nobels Explosive Co. Ltd. (Steventon and Dumfries), MQAD
Bishopton and MQAD Chorley.

REFERENCES

1. Alexander, W. J. and Mitchell, R. L., Anal. Chem., 21, 1497 (1949).

2. Segal, L., "Advances in Chromatography", Ed. Giddings, Vol.12, Chapter 2, Dekker (1975).

3. Badger, R. M. and Blaker, R. H., J. Phys. Colloid Chem., 53, 1056-1069 (1949).

4. Chang, M., Tappi, 55, (8), 1253-57 (1972).

5. Chromatix Application Note LS-1.

6. Chromatix Application Note LS-2.

7. Segal, L., J. Polym. Sci (Part C), No. 21, 267-82 (1968).

8. Carignen, Y. P. and Turngren, E. V., B. R. L. Report No. 1707, 405-434.

9. Segal, L., Timpa, F. D., Wadsworth, J. I., J. Polym. Sci., (A1), 8, 25-35 (1970).

10. Lindsley, C. H. and Frank, M. D., Ind. Eng. Chem., 45, 2491 (1953).

11. French, D. M., Naufleet, G. W., King, G. M., Report No. NOLX 81, Ordnance Laboratory, U. S. Navy, June 1974.

12. Grubisic, Z., Rempp, P., Benoit, H., J. Polym. Sci (Part B), 5, 753 (1967).

13. Jenkins, R. G., Masters Thesis, University of Waterloo, Canada.

14. Timpa, J. D., Segal, L., J. Polym. Sci., A9, 2099-2103 (1971).

15. Ouano, A. C., Barrall, E. M. II, Broido, A., Javier-Son, A. C., "Polymer Molecular Weight Methods", Ed. Ezrin, American Chemical Society., 1973., Chapter 17.

16. Spatorico, A. L. and Coulter, B., J. Polym. Sci., (Physics), 11, 1139 (1973).

17. Dawkins, J. V., Denyer, R. and Maddock, J. W., Polymer, 10 (3), 154 (1969).

GPC AND VISCOMETRIC INVESTIGATIONS ON GRAFTING REACTION OF SAN ONTO EPM ELASTOMER

A. De Chirico, S. Arrighetti and M. Bruzzone

Assoreni
Polymer Research Laboratories
San Donato, Milan, Italy

INTRODUCTION

Investigations on solution and bulk properties of grafted polymers are very often interesting for the novelty of physico-chemical topics that are usually connected with them. Solute/solvnet interaction[1], length and distribution of grafts[2],[3], phase separations are some problems concerning grafted polymers. Solution investigations frequently present a lot of difficulties for the low number of liquids in which grafted polymers are soluble. We want to report on some solution properties of a thermoplastic elastomer obtained by radical-induced grafting of styrene, S, and acrylonitrile, AN, on poly (ethylene-co-propylene), EPM, with 60% of ethylene. We have examined the influence of grafting reaction on starting EPM by GPC, the solution behaviour of some EPM-g-SAN samples by viscometry in mixed solvents and glass transition temperatures.

EXPERIMENTAL

GPC measurements were carried out with two Waters instruments, Models 200 and 244 using THF as elution liquid at a flow

rate of 1 ml/min. Intrinsic viscosities were determined from
measurements in mixed (vol/vol) solvent solutions at 30° C using an
Ubbelohde viscometer. Differential scanning calorimetry, DSC,
were performed with a Du Pont 900 thermal analyzer at a heating
rate of 10°/min.

RESULTS AND DISCUSSION

GPC Investigation

Resins[4] were obtained by grafting styrene and acrylonitrile (with S/AN = 3.17 by weight) on EPM and were extracted first
with hexane and then with methylethylketone, MEK, in order to
separate ungrafted EPM and SAN fractions from EPM-g-SAN samples.
GPC curves of starting and ungrafted EPM have shown that grafting
reaction of S and AN occurs on EPM fractions having higher molecular weights, MW.

Figure 1 shows plots of typical volume elution curves, of
starting, ungrafted EPM and corresponding grafted EPM-g-SAN. The
EPM curves have been normalized with respect to (ungrafted/starting)
EPM weight ratio, W_u[5]. Elution curve of ungrafted EPM fraction
corresponds to lower MW's of starting sample whereas the part of
elastomer with higher MW goes into EPM-g-SAN as backbone. Elution
curve of whole grafted sample reproduces the same shoulder at
higher MW's as the starting elastomer. Figure 2 shows elution
curves of a grafted sample obtained with a GPC (Waters 244) having
both refractive index and ultraviolet detectors. The latter
detector records phenyl group of SAN copolymer grafted onto EPM
backbone. SAN branch distribution of grafted EPM have been obtained
after calibration of the GPC. GPC elution curves of some ungrafted
EPM elastomers, normalized with respect to its own (ungrafted/star-
ting) weight ratio, are compared to that of the starting sample in
Figure 3.

All GPC curves of ungrafted EPM fractions shift towards lower
MW's as fast as the amount of grafted EPM increases but without
getting through elution curve of starting elastomer. According

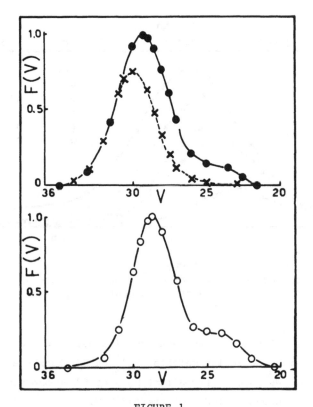

FIGURE 1

GPC Normalized Curves of Starting (●) and Ungrafted EPM (✗)
Extracted with Hexane from a Grafting Resin and of the Corresponding EPM-g-SAN (○).

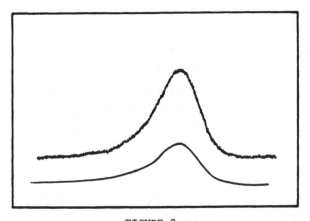

FIGURE 2

Ultraviolet (——) and Refractometric (⌇⌇⌇) GPC Traces of an EPM-
g-SAN Sample.

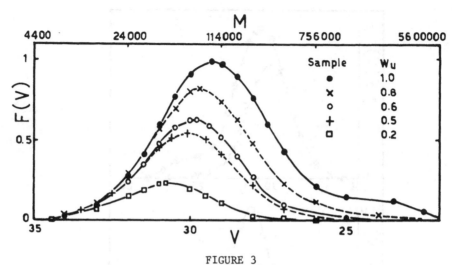

FIGURE 3

Trend of Normalized GPC Curves of Ungrafted EPM with Respect to Starting Sample as a Function of the Amount of Ungrafted EPM, w_u, in Resins.

to Collins et al.[5] we have calculated, by viscometric data, the viscometric-average molecular weights, $M_{v,g}$ of EPM backbones of some EPM-g-SAN. We have observed that $M_{v,g}$ of EPM-g-SAN samples is a function of amount, $1-W_u$, of grafted EPM.

DSC Investigation

It is known that grafted samples are able to display bulk properties of both components when there is no compatibility between them as well as SAN and EPM copolymers are. By DSC we have displayed the glass transition temperatures, Tg, of EPM-g-SAN samples. Experiments performed on a group of samples covering a range of SAN content from 10 to 60% have shown two Tg at 100° and -70°. These values correspond to the Tg's of SAN and EPM free copolymers and show how a phase separation between hard and soft dominions takes place in bulk.

Viscometric behaviour

Solution properties of EPM-g-SAN were further examined by viscometry in chosen mixed solvents. We have found many grafted samples are soluble in THF and chlorobenzene (CI-∅) and unsoluble, among other liquids, in isooctane (i-0) and in methylethylketone (MEK) which are good solvents for EPM and SAN respectively. Three grafted samples were chosen with increasing SAN/EPM ratio and equal viscosity average MW of EPM backbone $M_{v,g}$[5] (see Table 1) in order to compare the effect of SAN content on solution properties of grafted EPM elastomers.

Table 1 reports the intrinsic viscosities of SAN fractions extracted from resins with MEK (free SAN) as an indication of SAN chain length grafted onto EPM. Some results reported in literature[2,3] actually show that polymers grafted onto a backbone and ungrafted ones (free polymers) have equal MW. Intrinsic viscosities measured in CI-∅/i-0 and CI-∅/MEK mixtures for three grafted EPM samples are plotted in Figure 4. We have observed different viscometric behaviours connected with graft lengths and SAN contents of the samples. We can infer that a change of chain conformation occurs passing from an elastomeric coil with a few and short grafts such as from sample A to sample C which is only able to dissolve in mixtures containing MEK because of its long SAN grafts.

TABLE 1

Data concerning three EPM-g-SAN samples having a $M_{v,g}$ of 340,000.

Sample	$\dfrac{SAN}{EPM}$, by weight	$[\eta]$ 30° MEK of free SAN	$[\eta]$ 30° THF
A	0.25	0.4	3.2
B	1.10	0.5	2.5
C	1.04	1.5	1.7

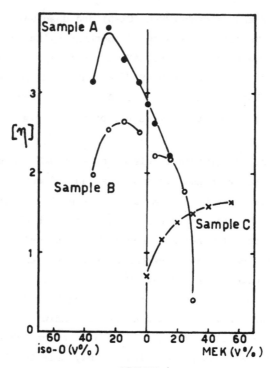

FIGURE 4
Intrinsic Viscosities of Some EPM-g-SAN Samples in Mixtures of
Chlorobenzene/MEK and Chlorobenzene/Isooctane.

REFERENCES

1. Dondos, A., Patterson, D., J. Polym. Sci. Part A2, 5, 230,
    (1967).

2. Locatelli, J. L., Riess, G., Angew. Makromol. Chem., 28, 161
    (1973)

3. Kuhn, R., Alberts, H., Bartl, H., Makromol. Chem. 175, 147
    (1974)

4. Arrighetti, S., et al., (Snamprogetti S.p.A.), Italian Patent
    22543 (1976); Chem. Abst., 87, 23983 h (1977).

5. Collins, R., Huglin, M. B., Richards, R. W., Eur. Polym. J.
    11, 197 (1975).

GPC OF POLYAMIDES

D.J. Goedhart, J.B. Hussem and B.P.M. Smeets

Akzo Corporate Research Arnhem
Arnhem, Velperweg, The Netherlands

ABSTRACT

A critical survey is given of all solvents for aliphatic polyamides which have been used for GPC analysis. The experience we
have gained with four of these solvents is not very encouraging.
Some results are given of measurements with hexafluoroisopropanol
and µBondagel® columns. Also a new solvent is introduced, namely
m-cresol/dichloromethane (1/9). This mixture can be used with
µStyragel® columns at low temperature and it is suitable for polystyrene, polyesters, polyamide 6 and 6.6.

INTRODUCTION

Several solvents have been proposed for the analysis of
aliphatic polyamides. Solvents at moderate temperature are

-strong acids, e.g. sulfuric acid, formic acid

-phenol derivates, e.g. m-cresol, o-chlorophenol

-fluorinated alcohols

-alcoholic salt solutions, e.g. methanol + $CaCl_2$

-hexamethylphosphorotriamide.

Not all these solvents are quite suitable for GPC. An ideal
solvent for GPC has the following properties:

-low viscosity

-low (or very high) refractive index

-translucent for UV

-non-corrosive

-non-toxic

-oxidative and thermal stable

-usable at room temperature

-solvent for polystyrene and other polymers

-compatible with the column packing

-no association or degradation of the polymer in that solvent

-known Mark-Houwink parameters of the polymer and the cali-
    bration standards.

It is obvious that none of these solvents can meet all requirements and with every choice some draw-backs have to be taken
for granted. Until now we have used 6 different solvents for GPC
analysis of polyamides; none of them proved to be ideal. In the
following section, a survey is given of the solvents mentioned in
the literature and our experiences with some of them.

APPLIED SOLVENTS FOR GPC OF POLYAMIDES

First mention of GPC of polyamides in the literature is by
Goebel[1]. He used m-cresol at 135° with Styragel® columns for
PA-6 and PA-6.6. Dudley[2] used m-cresol at 130°C for PA-6.6 and
considered it as an ideal solvent. On the other hand, Ede[3] found
that PA-6 degraded considerably in m-cresol at 130°C within a few
hours. He therefore proposed the use of a 1:1 (v/v) mixture of m-
cresol and chlorobenzene at 43°C. This mixture has a far lower viscosity than pure m-cresol and, from relative viscosity measurements,
Ede concluded that it is even a better solvent for PA-6. Reproducible results could be obtained, proveded that 0.25% benzoic acid was
added to prevent adsorption effects. At our laboratory we have
used this solvent at room temperature for the analysis of PA-6,
PA-6.6, PA-6.10. PA-11 gives a very low peak and PA-12 gives no
peak at all because of the very small dn/dc. At first it looked
promising because polyetheneterephthalate (PETP) is also soluble in

this mixture. However, we found it rather difficult to get reproducible results due to base-line instability. Slight changes in
the solvent composition are easily detected by the differential
refractometer. These changes are difficult to avoid, as the
vapour pressure of chlorobenzene is much higher than that of m-
cresol. The drawing of the base-line is also hindered by the large
vacancy peaks which always occur when working with solvent mixtures. These peaks, at the low molecular weight end of the chromatograms are caused by preferential adsorption of m-cresol by the
polymer. This adsorption is an advantage when light scattering
measurements are made, for, on a molecular scale, the dn/dc is
larger than the dn/dc measured in the conventional way [4,5,6].
More troublesome is the fact that the adsorption is dependent
upon the molecular weight and upon the concentration[7,8]. This
effect distorts the chromatograms and makes it more difficult to
obtain the true molecular weight distribution.

A similar system is mentioned by Ishida and Kawai[9]. They
used m-cresol and chlorobenzene in the ratio 1:3.5 (v/v) with o.3%
benzoic acid for PA-6 and PA-6.6 and a mixture of m-cresol and
chloroform for PETP. A mixture of o-chlorophenol and chloroform
1:3 (v/v) at room temperature, suitable for polyamides and polyesters, has been proposed also[10]. The polymer is first dissolved
in pure o-chlorophenol with subsequent dilution with chloroform. At
our laboratory we have also tried this solvent. With PETP we got
good results but with PA-6 we never observed even a trace of a peak.
According to Cazes[11] a peak appears after a very careful purifica-
tion of the o-chlorophenol. No explanation can be given of this
strange phenomenon. Walsh[12] used pure o-chlorophenol at 100°C
for PA-6. He found adsorption effects, a decrease in molecular
weight of 3%/h and a large difference between the hydro-dynamic
volume calibration curves of polystyrene and polyamide.

A new approach of the problem came from Panaris and Pallas[13]
They used hexamethylphosphorotriamide (HMPT) at 85°C for PA-6, PA-11
and PA-12. The advantages of this solvent are its relatively low
refractive index and viscosity, and it is a powerful solvent for many

other polymers such as PVA, PVC, PMMA, PS. No adsorption effects
were found with PA-11. At our laboratory, we have used this solvent for the analysis of PA-6, polyvinylpyrrolidone and polyvinyl-
idenefluoride. Initial experiments were encouraging. However,
some batches of HMPT were unstable. Time and again, an insoluble
gel was formed, which blocked pump and columns. As it also turned
out that HMPT is potentially carcinogenic[14], we decided to ter-
minate all work with it. We then changed to dimethylacetamide +
2.5% LiCl at 80°C. Polystyrene and PA-6 could be dissolved at
about 100°C, but PA-6.6 dissolves only above 130°C. Injections
of PS gave a positive LiCl-peak and with PA-6 a negative peak
appears which interferes with the low molecular weight part of the
chromatogram. The negative peak could be diminished by adding an
extra 0.1% LiCl to the 1% PA-6 solution. However, stainless steel
316 slowly corrodes in this solvent, resulting in repeated leakage
in the tubing system. Consequently, this solvent cannot be
recommended.

Attractive but expensive solvents for polyamides are 2,2,2-
trifluoroethanol (TFE) and 1,1,1,3,3,3-hexafluoropropanol-2 (HFP).
Dudley (2) used TFE at 30°C with Styragel® columns but was not
successful as his columns degraded within a few weeks. This is
probably due to the use of a 250 Å column. Styragel® columns with
pore sizes less than 1000 Å should not be used with solvents which
are non-solvents for polystyrene, according to Waters Associates.
Provder, et al.,[15] were more successful with TFE at 50°C Styragel®
columns, but their smallest pore size was 15000 Å. Since PS is
not soluble in TFE they had to use a rather elaborate method for
obtaining a calibration curve. It is remarkable that they do not
mention a polyelectrolyte effect in this solvent. Matzner, et al.,[16]
also used pure TFE with Styragel® columns at room temperature. Their
column set contained one with 250 Å pore size, apparently without
any problems, contrary to the experiences of Dudley[2].

All experiments mentioned above were carried out with large
Styragel columns and GPC model 200 of Waters Associates which con-

sumes a large amount of solvent. Modern HPGPC equipment with
μStyragel® or μBondagel® columns drastically reduces the solvent
consumption, making the use of expensive solvents like TFE and
HFP mre feasible. Successful determinations of PA-6, PA6.6 and
PA-6.10 in TFE with μStyragel® columns are reported by Dark[17].
μStyragel® columns have been used by Drott[18] for 2 years with
HFP at 25°C without any degradation of the columns. The polyelectrolyte effect of PA-6.6 could be suppressed by the addition
of 0.01 M sodium trifluoroacetate (NatFAc). On the other hand,
Waters Associates mention the use of μStyragel® with HFP for the
analysis of PA-6, apparently without any salt addition[19].
Chromatography of a polyester on a μBondagel® column with HFP
is reported by Vivilecchia, et al[20].

   We want to report on the chromatography of polyamides with
HFP on μBondagel® columns. These solvents have been chosen because they are both suitable for polyamides and polyesters.
Solution experiments indicated that similar measurements could
have been done with chloroform instead of dichloromethane. We
preferred the latter because it is less toxic, more stable and its
refractive index is lower.

EXPERIMENTAL

Apparatus

   The chromatograph consists of components of different manu-
factures, viz. Varian 8500 pump, a pneumatically operated Rheodyne
injection valve model 70-10, Knauer column oven type 6000, Knauer
UV/RI dual detector type 6100 with 10 μl cell, Haake water bath
F_3 - S for temperature control of the dual detector, Chromatix
KMX-6 low angle laser light scattering photometer (LLALS) with
temperature controlled GPC cell. Columns were μBondagel® and
μStyragel® from Waters Associates. Several filters were used in
order to prevent blockage of the columns and to reduce the noise
of the light scattering detector. The solvents were filtered
through a 0.2 μm Teflon filter (Millipore), the solutions through

a 0.9 μm Versapor filter (Gelman) and a 0.5 μm Teflon filter.
The columns were further protected by a 2 μm Waters precolumn
filter and a 0.2 μm Teflon filter was placed after the columns.

The signals of the three detectors and the pump pressure were
registered with a 4-pen Watanabe recorder type MC 641. The chromatograms were digitized with a "pencil follower" of D-Mac Ltd. with
crosswire viewer, paper tape output and a resolution of 0.1 mm.
At least 100 datapoints per chromatogram were taken. Calculations
and plotting were carried out with the aid of a computer.

Operating conditions

The solvents are degassed by applying vacuum and stirring for
a little while and are then stored in the solvent reservoir under
helium pressure. HFP measurements have been carried out with one
μBondagel® E-linear column with nominal molecular weight separation
range 2,000 – 2,000,000, a 20 μl sample loop, flow rate of 0.25
ml/min, all temperature controls set at 40°C and chart speed at
3 cm/min. The pressure drop was approximately 20 bar. Column performance was checked, previously, with toluene in dichloromethane
with a flow rate of 1 ml/min, a 10 μl sample loop and temperature
30°C. With the tangent method we obtained a plate count of 5200,
with the 5σ method 4430. After conversion to HFP these numbers
were 3780 and 2500, respectively. Measurements with m-cresol/di-
chloromethane have been carried out with a set of 5 μStyragel®
columns with nominal exclusion limits of 10^4, 10^5, 10^3, 500 and
100 Å, respectively. Sample loop was 200 μl, flow rate 1 ml/min,
all temperature controls set at 30°C and chart speed at 0.5 cm/min.
Pressure drop was approximately 35 bar.

Solvents

HFP was obtained from E.Merck, Darmstadt (Uvasol® quality)
and redistilled after use. Sodium trifluoroacetate was made from
trifluoroacetic acid and sodium hydroxide, precipitated with 1,4-
dioxane and recrystallized from ethanol[21]. Dichloromethane was
a Baker Analyzed Reagent, m-cresol was a technical grade, purified
by vacuum distillation under nitrogen.

Solvents

HFP was obtained from E. Merck, Darmstadt (Uvasol® quality) and redistilled after use. Sodium trifluoroacetate was made from trifluoroacetic acid and sodium hydroxide, precipitated with 1,4-dioxane and recrystallized from ethanol[21]. Dichloromethane was a Baker Analyzed Reagent, m-cresol was a technical grade, purified by vacuum distillation under nitrogen.

Solutions

PA-6, PA-6.6, PA-12 and moderately crystalling PETP are easily dissolved in HFP at room temperature. Highly crystalline PETP has first to be dissolved in e.g. phenol/1,1,2,2-tetrachloroethane 60/40 (w/w) at 100°C or higher[22]. The polymers do not dissolve directly in the mixture m-cresol/dichloromethane. First a solution is made of 0.1 g polymer in 2 ml m-cresol, which is then diluted with 18 ml dichloromethane. The solutions remain stable for at least a few days.

RESULTS WITH HFP/ μBONDAGEL®

Two polyamide-6 samples L6 and H6 have been used to test this solvent-column combination. The whole polymers have been characterized with light scattering and dilute solution viscosity, both at 40°C in pure HFP. The results are given in Table 1.

The light scattering measurements have been carried out with the KMX-6, using the GPC flow-through cell. All operating conditions were the same as with the measurements of the GPC fractions, except that the solutions were pressed through the cell by means

TABLE 1

Solution Properties of Polyamide-6 in pure HFP at 40°C.

	L6	H6	units
Mw	33700	58000	dalton
A_2	7.98	4.80	mol cc/g^2 x 10^3
[η]	1.171	1.56	dl/g
k	0.110	0.106	–

of a simple syringe pump. The intrinsic viscosities have been obtained with the Kraemer equation

$$\frac{\ln\eta_r}{c} = [\eta] - k[\eta]^2 c$$

The results are not in agreement with those of Pavlov et al[23], who found the Mark-Houwink parameters K = 0.0048 and a = 0.65 if HFP + 0.05 M NatFac at 20°C. For instance the relation of Pavlov gives [η] = 6.0 for Mw = 58100. Nevertheless, we found a fair agreement between the molecular weights of the fractions measured with the LALLS and thos calculated from the hydrodynamic volume calibration curve and the Pavlov relation as is shown in Figure 1. No explanation can be given for this contradiction up till now.

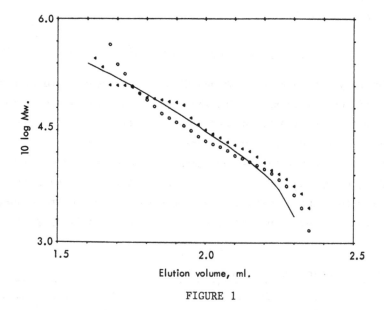

FIGURE 1

M_w versus Elution Volume (-from PS Hydrod. Volume Calibration Curve and Mark-Houwink Relation of PA-6 in HFP, O Fractions of H6, ▲ Fractions of L6).

RESULTS WITH M-CRESOL/DICHLOROMETHANE

With this solvent mixture we always get rather large vacancy peaks, but they appear later than the peak of the cyclic timer, so they do not disturb the chromatogram. In Figure 2 the chromatograms are given of a PA-6 sample, one before and one after a heat treatment with relative viscosities in m-cresol of 2.51 and 1.64, respectively. The degraded sample shows a more pronounced high molecular weight shoulder. Up till now it was not possible to get good light scattering results due to the relatively high scattering of the solvent and the presence of dust. New experiments are planned with more thoroughly cleaned solvent.

CONCLUSION

It has been shown that many solvents of polyamides are not quite suitable for GPC analysis. Hexafluoroisopropanol with a salt added and trifluoroethanol have certain advantages. Drawbacks

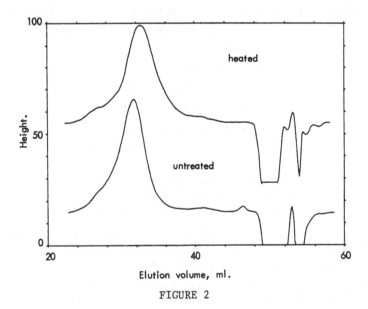

FIGURE 2

Sample PA-6:   Before and After Heat Treatment.

are the high price and that μBondagel columns with small pores are
not available and it is doubtful if μStyragel with smaller pores
than 1000 Å can be used. The new solvent mixture m-cresol/di-
chloromethane looks very promising. Advantages are that the calibration can be carried out directly with polystyrenes and gel with
small pores can be used, allowing a better resolution in the
oligomer range.

REFERENCES

1. Goebel, C. V., 4th Int. Seminar GPC, Miami Beach 1967.

2. Dudley, M. A., J. Appl. Polym. Sci. 16, 493 (1972).

3. Ede, P. S., J. Chromatogr. Sci. 9, 275 (1971).

4. Tuzar, Z., Kratochvil, P., Bohdanecky, M., Adv. Polym. Sci.
    30, 117 (1979).

5. Tuzar, Z., Kratochvil, P., Macromolecules 10, 1108 (1977).

6. Huglin, M. B., Richards, R. W., Polymer 17, 587 (1976).

7. Dondos, A., Benoit, H., Int. J. Polymeric Mat. 4, 175, (1976).

8. Hert, M., Strazielle, C., Int. J. Polymeric Mat. 4, 195, (1976).

9. Ishida, Y., Kawai, K., Shimadzu Hyoron 29112, 89 (1972).

10. Water Assoc., P/P technical note 7 (11-4-76).

11. Cazes, J., private communication.

12. Walsh, E. K., J. Chromatogr. 55, 193 (1971).

13. Panaris, R., Pallas, G., J. Polym. Sci. B8, 441 (1970).

14. Zapp, J. A., Jr., Science 190, 442 (1975).

15. Provder, T., Woodbrey, J. C., Clark, J. H., Separ. Sci. 6,
    101 (1971).

16. Matzner, M., Roberson, L. M., Greff, R. J., McGrath, J. E.,
    Angew. Makromol. Chem. 26, 137 (1972).

17. Dark, W. A., Appl. note Waters Assoc., AH 352, jap. 1975.

18. Drott., E. E., Proc. Int. Symp. Liq. Chromat. Anal. Polymers
    1976 Houston (Chromatogr. Sci. Series 8, 41).

19. Waters Assoc., P/P technical note 13 (1-18-77).

20. Vivilecchia, R., Lightbody, B., Thimot, N., Quinn, H., Proc.
    Int. Symp. Liq. Chromat. Anal. Polymers 1976 Houston (Chroma-
    togr. Sci. Series 8, 11).

21. Auerbach, I., Verhoek, F. H., Henne, A. L., J. Am. Chem. Soc.
     72, 299 (1950).

22. DIN 53728, Part 3.

23. Pavlov, A. V., Bresler, S. Ye., Rafikov, S. R., Vysokomol.
     soyed. 6, 2068 (1964).

CONTRIBUTION OF GEL PERMEATION CHROMATOGRAPHY (GPC) TO THE CHARACTERIZATION OF ASPHALTS

B. Brulé

Laboratoire Central des Ponts et Chaussées
Paris, France

INTRODUCTION

A better understanding of the structure of asphalts is desirable, not only for scientific purposes, but also from the technological and economic points of view as this binder is used in road technology in France, to the extent of about 3 million tons annually. Assuming that a road covering contains less than 10% of asphalt, more than 30,000,000 tons of material, used annually to lay foundations and driving surfaces of roads and highways, depends on the properties of the binder for its performance and longevity.

Research workers therefore seek a full understanding of the colloidal structure of this binder, as this would make it possible to relate compositional parameters to actual properties. Among the physico-chemical analytical techniques which could be useful in the characterization of such a medium, gel permeation chromatography (GPC) is a good choice because of its efficiency and ease of application. Further, the recent development of supports with small particle size has added yet another advantage, namely a response time which is of the same order of magnitude as that attained in gas-liquid chromatography. In view of this situation, it was of interest to carry out a systematic study. The following report gives an

account of the application of GPC on micro-packing to the characterization of asphaltic binders.

The experiments carried out involved analysis of nine asphalts
with appreciably different degrees of penetration and manufacturing
processes. We first show that GPC makes it possible to distinguish
the various samples qualitatively within ten minutes or so. Second-
ly, observation of the modification of the chromatograms as a function
of operating parameters, such as flow rate or quantity injected, is
interpreted in terms of dissociation of micellar agglomerates. Finally, we suggest a quantitative interpretation of the chromtograms involving the experimental construction of, on the one hand, correction
curves for the detectors, and on the other a calibration curve of
molecular weight which is specific for asphalts.

FRACTIONATION AND CHARACTERIZATION OF ASPHALTS BY GPC

GPC was used in the study of heavy petroleum fractions by Altgelt
[1,2] as early as 1965, i.e., very soon after the "invention" of polystyrene gels crosslinked with divinylbenzene. Tests on asphaltenes
and maltenes led to estimation of the lower limit of molecular weight
for asphaltenes at about 700 (the upper limit reaching about 40,000),
whereas maltenes contain constituents for which the molecular weight
may reach 24,000. This shows the lack of selectivity in the asphal-
tene-maltene fractionation.

Breen and Stephens[3] attempted to establish a correlation between the glass transition temperatures of about fifty samples of
asphalts and their molecular size distributions, but without arriving
at any conclusive result. Richman observed qualitative differences
in the GPC chromatograms of different asphalts[4], especially toward
the higher molecular weights.

Bynum and Traxler[5] compared the GPC chromatograms of some
samples of road asphalts with variable practical qualities (some being considered as excellent, some as average and others as poor), and
observed qualitative differences in the shape of the chromatograms
as a function of the origin of the crude.

A preparative apparatus, which permits fractionation of 1 to 10
g of material, was used by Albaugh et al.[6] for experimental construction of calibration curves for distillation residues from petroleum crudes. It is interesting to note, according to Albaugh, that
although the charge in the columns (between 1 and 10 g) affects the
shape of the preparative chromatogram (molecular weight of fraction
as a function of elution volume), its effect on the polydispersity of
the fractions is negligible. Finally, a considerable difference was
observed between the shape of the chromatogram recorded with the differential refractometer and that of the chromatogram constructed by
weighing the fractions.

Dickson et al.[7,8] fractionated and analysed a distillation
residue from Kuwait crude under conditions identical to those of
Albaugh[6], and compared the calibration curve of molecular weight
determined by vapor pressure osmometry, with a curve giving the relationship between elution volume and the molecular weight of the
"elementary structural sheet" in the Yen[9] model such as may be determined by N.M.R. Agreement is satisfactory for small molecules,
but the curves are considerably different from each other for molecular weights greater than about 1000.

Snyder has studied the problems presented by the application of
analytical GPC to the characterization of asphalt[10]. In particular,
he examined the response of the refractometric detector and of a UV
detector at 370 nm as a function of elution volume and showed that,
unlike the case of polymers, this response was not independent of
elution volume.

Reerink and Lijzenga[11] suggest a different method for calibration of GPC based on comparison with results obtained by ultracentrifugation. The principle of the method consists of constructing the
integral curve of molecular weight distribution as a function of
elution volume by GPC, and of determining the same integral curve of
molecular weight distribution experimentally but, this time, as a
function of the molecular weight obtained by ultracentrifugation. A
comparison of these two curves makes it possible to define the relationship between elution volume and molecular weight.

From this examination of the literature concerning the application of GPC to the characterization of bituminous compounds, it is evident that much research work has been carried out on the subject but its fragmentary nature would suggest it raises more problems than it solves. In fact, although an attempt has been made to resolve most of the special difficulties presented by the use of this technique with bituminous binders, i.e. problems of detector response, of calibration and of the ideal or non-ideal nature of solutions depending on association-dissociation phenomena related to the concentration, the answers are often contradictory.

Finally, we may note that no experiments in the field of asphalts have been carried out to date using micro-supports, of which the advantages (low volume and high elution rate, therefore very low response time compared to traditional supports), are now well known.

EXPERIMENTAL STUDY OF OPERATING PARAMETERS IN QUALITATIVE GPC

This phase of the study is based on the qualitative comparison of chromatograms obtained under given operating conditions. Construction of a calibration curve of molecular length using polystyrene standards and normal alkanes with 6 to 32 carbon atoms, makes it possible to associate a molecular size with each distribution coefficient. It is clear that such a dimension is defined only by the experimental procedure which measures it, but it is nevertheless a useful and reproducible label. The ordinate in this case corresponds to the response of the detector.

Choice of Columns and Detectors

Experiments were carried out with two systems of μStyragel Columns:

set of four columns: 10^3, 10^4, 10^5, and 10^6 $\overset{o}{A}$
set of two columns only: 10^3 and 10^4 $\overset{o}{A}$

Figure 1 shows the chromatograms of three CFR asphalts and of the Esso Port Jerome asphalt, obtained with the first system of columns for an injected quantity of 2 mg in THF at a flow rate of 2.3 ml/min, detection being by means of differential refractometer.

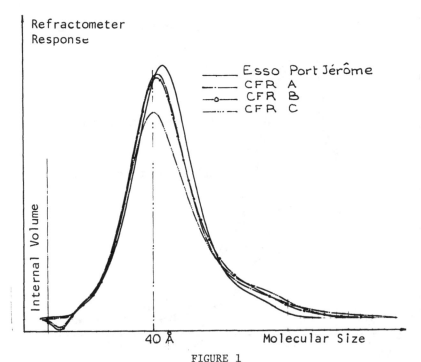

FIGURE 1

Chromatograms Obtained by Differential Refractometry of 180/220
Asphalts, 2 mg inj., 4 Columns µStyragel.

It is seen that, under these conditions, any differences ther may
be between these various types of asphalts are not clearly defined.

The asphalts were re-injected under identical conditions as to
solvent and flow rate, the quantity of test sample being reduced to
0.125 mg, and the refractometric detection being supplemented by UV
detection at 254 nm with an optical path of 10 mm. The chromatograms
are reproduced in Figure 2. Although the differences are more dis-
cernible than in the case of Figure 1, they are still no less difficult to interpret.

In view of the dispersion of molecular sizes observed with bituminous binders, the use of columns of high porosity (10^5 and 10^6 Å)
did not seem to offer any special advantage. We therefore decided
to remove them and work only with two columns. Also, in order to
make the UV detection more specific for polycondensed compounds, we

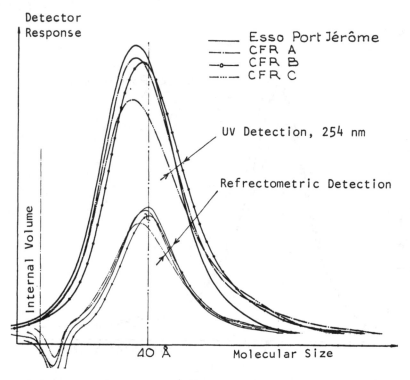

FIGURE 2
Chromatograms of 180/220 Asphalts for 0.125 mg inj., 4 Columns μStyra-gel.

changed the wavelength and set it on 350 nm. Finally, in order to compensate for the large difference of response coefficients between refractometry and ultraviolet absorption, we reduced the optical path of the cell to 3 mm.

Under these new conditions, and for a THF flow rate of 3.5 ml/min, the chromatograms obtained for the five 180/220 asphalts are as reproduced in Figure 3. These experiments show that any slight differences there may be between various asphalts with the same penetration are difficult to demonstrate by refractometric detection or by UV detection at 254 nm. We may conclude from this that, at least in the case of asphalts with high penetration, the differences are not essentially concerned with the molecular weight distribution of the constituents (refractometric detector) nor with the distribution of

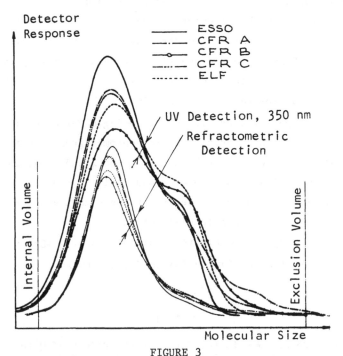

FIGURE 3

Chromatograms (Refract. and UV @ 350 nm) of 180/220 Asphalts, 1 mg inj., 2 Columns μStyragel of 10^3 and 10^4 Å.

aromaticity (UV detection at 254 nm); but rather with the relative content of polycondensed compounds such as amy be revealed by ultra-violet absorbance at 350 nm. We shall therefore retain the "classi-cal" conditions for GPC examination of asphaltic substances, i.e. two columns of μStyragel of 10^3 and 10^4 Å, and double detection by differential refractometry and measurement of optical density at 350 nm.

Effect of Quantity Injected

The choice of columns and detectors having been fixed, it was now important to study the effect, on the shape of the chromatograms, of the quantity injected. It may happen that this quantity is limit-ed at an upper level, either by "overload" of the column or by sat-uration of the one or more detectors. Its lower level is fixed by

the minimum response levels of the detectors, and shall be the lower
the more sensitive is the detector. Figure 4 shows the chromatograms
obtained by differential refractometry and UV, of the asphalt Elf
Grandpuits 80/100 for injected quantities varying from 4 mg to 0.015
mg, the operating conditions being given in Table 1. The solvent
used was THF at a flow rate of 3.5 ml/min.

Examination of Figure 4 leads to the following observations:

-judging from the superposition of the chromatograms obtained
by refractometry for injected quantities of 2 and 4 mg, the
latter value is less than the overload limit of the columns.
This limit would lead to a widening of the signal and its
shift towards the lower molecular weights.

-it is practically impossible to obtain a workable signal
from the refractometer for an injected quantity less than
0.25 mg.

TABLE 1

Experimental Conditions in the Study of the Effect of Quantity Injected on the Shape of the Chromatograms.

Volume Injected (µl)	Concentration (%)	Quantity Injected (mg)	Sensitivity of Detector	
			Refractometer	UV @ 350nm
40	10	4	32X	N.L.*
20	10	2	16X	N.L.*
20	5	1	8X	0.64
20	2.5	0.5	4X	0.32
20	1.25	0.25	2X	0.16
20	0.62	0.12	-	0.08
20	0.31	0.062	-	0.04
20	0.16	0.031	-	0.02
20	0.08	0.016	-	0.01

*The position of the sensitivity selector corresponds to a region in
which the detector is non-linear.

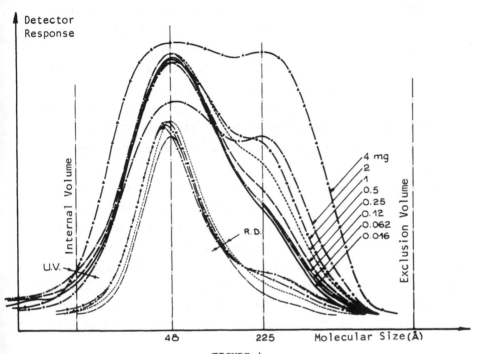

FIGURE 4
Effect of Quantity Injected on the Shape of the Chromatogram of an
Elf Grandpuits 80/100 Asphalt (the Chromatogram for Injected Quantity
of 0.031 mg is Superimposable on that Obtained for 0.016 mg).

-the UV detector may be used, under normal conditions (region
in which the response is linearly related to the concentration), only for test samples less than or equal to 1 mg.
-its high sensitivity and stability make it possible to obtain a workable graph for injected quantities as low as
0.016 mg.
-whatever be the type of detection the chromatogram depends on
the quantity injected, decrease of the latter leading to a
reduction in the amount of compounds of large molecular size.
We may conclude that the GPC examination of asphalts with double
detection (refractometry and UV at 350 nm) is possible for injected
quantities from 1 to 0.25 mg. Below this value down to 0.016 mg,
only the response of the UV detector can be utilized.

We know that asphalt is a complex medium with colloidal structure. The passage of such a substance into solution does not necessarily lead to elimination of all the intermolecular interactions
responsible for the formation of micelles and agglomerates. It may
therefore be considered that the change in shape of the chromatograms
as a function of quantity injected is due to the effect of dilution
on the dissociation, this latter making itself especially felt between the concentrations 10% and 0.62%, and seeming to disappear
completely near 0.16%.

The "normal" condition for examination of an asphalt, therefore,
consists of injecting 1 mg of the substance. The dissociation process may be shown by decreasing the concentration of the solution
injected by a factor up to 16. It seems important to note that,
although decrease in the concentration of the solution injected
appears to lead to a completely dissociated state, its increase, under conditions compatible with the methods used, leads to observing
the binder in a state of association which is nearer the real situation.

Effect of Solvent

The tests mentioned above show that the shape of the GPC chromatogram of an asphalt depends on the quantity injected because of the
dissociability of certain constituent species. We then wished to
demonstrate the possible role of the nature of the solvnet, and to
examine what effect it might have on the dissociation at a given
concentration. Also, the presence or absence of a chromatographic
"drag" should enable us to form an idea of the interactions liable
to develop between the material analysed and the support.

Four good solvents for asphalts of different structure and dielectric constant, were chosen--THF, chloroform, benzonitrile and
tetrahydronaphthalene (tetraline). Benzonitrile and tetraline were
used as received, whereas THF was distilled shortly before use.
Chloroform was destablized by fractional distillation and then treated with activated silica gel.

Experiments were carried out using the asphalt Shell 40/50 and
Figure 5 shows the four chromatograms. We may add that, in order to
make these chromatograms rigorously compatible (elimination of fluc-
tuations of internal volume and of response coefficient of the UV
detector as a function of the nature of the solvent), they were
normalized using an electronic calculator equipped with a curve plot-
ter.

It is seen that the picture of molecular size distribution depends very strongly on the nature of the solvent, as the apparent
concentration of compounds with large molecules (third population of
average value approx. 800 Å) decreases appreciably in the order
tetraline, THF, chloroform, benzonitrile. This observation confirms
the hypothesis of the dissociability of certain species, the latter
depending not only on the concentration but also on the nature of the
solvent. We also note that the dissociating power of the solvent is

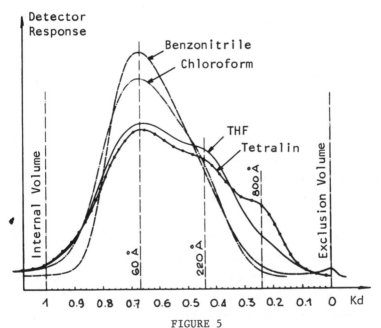

FIGURE 5
Shape of Chromatogram of a 40/50 Asphalt as a Function of the Nature
of the Solvent.

not only a function of its dielectric constant because, although the solvent with the lowest dielectric constant (tetraline: $\varepsilon = 2.757$) effectively corresponds to maximum association, and that with the highest dielectric constant (benzonitrile: $\varepsilon = 25.2$) corresponds to the most dissociated form, an inversion is observed for THF ($\varepsilon = 7.25$) and chloroform ($\varepsilon = 4.806$) for which the solubility parameters are, moreover, identical (9:1). This implies that the interactions responsible for association of the asphaltic constituents are not only of electrical nature, but also bring in more complex structural parameters. Finally, we note that only benzonitrile reduces the chromatographic drag observed in the region of lower molecular weights with the three other solvents.

The objective is the characterization, by GPC of bituminous compounds in the maximum state of association compatible with the technique. This would lead to carrying out the analysis in tetraline. However, because of the disadvantages associated with this solvent, in particular its limited solvent power and high viscosity, we decided to employ THF routinely.

Effect of Flow Rate

The effect of flow rate on the shape of the chromatogram of an asphalt was studied with two 40/50 asphalts (Shell and CFR). The quantity injected was 1 mg, detection was by means of UV, and the flow rate was varied from 0.6 to 7 ml/min (Fig. 6). It is seen that, although the flow rate does not much affect the positions of the average values of the various populations, its effect is not negligible on the apparent amount of constituent elements with large molecules.

Study of the effect of concentration of the injected solution has shown that the constituents with large molecules are dissociable. The kinetics of dissociation may be the cause of the decrease in the apparent content of agglomerates. We inject, in fact, 20 µl of solution of a sample eluted by about 10 ml of solvent, which corresponds to a dilution coefficient inside the apparatus of the order of 500. This dilution takes place in 3 1/2 min at a flow rate of 7 ml/min,

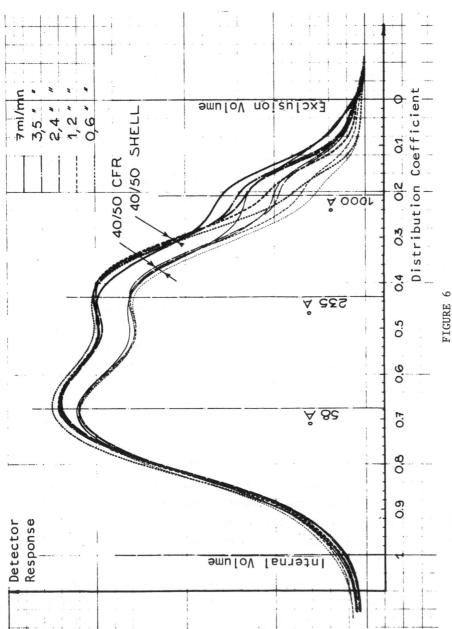

FIGURE 6

Effect of Flow Rate on Shape of Chromatograms of 40/50 Asphalts.

and in 40 min at a flow rate of 0.6 ml/min. If the kinetics of dissociation is relatively slow, it is then logical that we observe the
substance to be more associated the higher the flow rate.

Effect of Age of the Solution

We examined the effect of the time lapse between preparation of
the solution and its injection. Figure 7 shows the chromatograms of
two 40/50 asphalts for an injection of 1 mg with UV detection, the
solution having been prepared 4 hours, 7 days, 14 days and 21 days
before injection. There is an appreciable change in the chromatogram
between 4 hours and 7 days, this being towards a decrease in the population centered on the inferior molecular sizes together with an increase in the two others. There is then a quasi-stabilization of the
phenomenon.

Discussion of Study of Operating Parameters

At this stage, it is possible for us to make some comments.
Firstly, we have been able to establish that, to obtain a precise and
reproducible GPC picture of the molecular size distribution in bituminous binders, care must be taken to fix all the operating parameters
very rigorously. Then, we have seen that this picture varied from
that of a relatively associated state for high concentrations and
high elution rates, to that of a more dissociated state of solutions
of low concentration, very low elution rates or a very polar solvent
such as benzonitrile were used. One of the observations which should
be made is that, under the operating conditions adopted, i.e. THF at
a flow rate of 3.5 ml/min, there is a bimodal distribution of molecular sizes in samples of asphalts manufactured by direct distillation.
This distribution systematically becomes a trimodal in the case of
products having been subjected to blowing at some stage of their
manufacture. From this we conclude that, as the entities constituting the third population of greatest size generated by blowing are
dissociable, the operation has the effect of <u>creating active sites</u>
<u>giving rise to specific interactions favoring the formation of large</u>
<u>aggregates</u>.

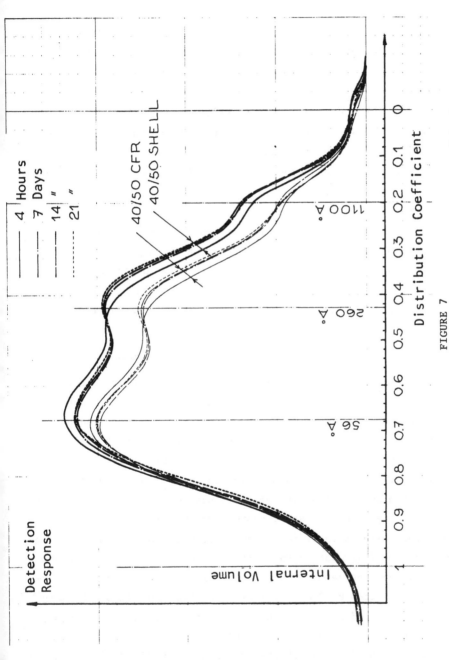

FIGURE 7

Change in Chromatograms as a Function of Time Lapse Between Solution
Preparation and its Injection.

In fact, while the operating parameters in GPC do have a considerable influence on the molecular size distribution picture in asphalts, it is because of the particularly unstable nature of the solutions. These are not ideal in the thermodynamic sense of the term but are micellar, the tern micellar being taken in its widest meaning.

The first aspect of the special behavior of these solutions is the reduction in the apparent concentration of compounds with large molecules, observed when the concentration of the solution injected is decreased. It may be deduced that dilution causes dissociation of the micellar agglomerates.

Interpretation of the change in the shape of the chromatograms with time is more complicated since it implies a rearrangement with very slow kinetics. The first hypothesis which comes to mind is that of the superposition of two phenomena. The first would be the dissociation of certain agglomerates during the process of solutions, and the second the slow reassociation of reactive species liberated by the first phenomenon.

QUANTITATIVE INTERPRETATION OF THE CHROMATOGRAM

There are two problems presented by the quantitative interpretation of the chromatograms. The first is very common in GPC when we are applying it to substances of low molecular weight. Actually, in this field, the question which arises first is that of calibration. The best method consists of calibrating the system in a specific way by means of standards of which the nature and chemical structure are dientical to that of the substance to be characterized. If these standards do not exist, recourse must be had to classical procedures for calibration of molecular size, molecular volume or hydrodynamic volume (the most suitable method must however be chosen and the choice does not seem obvious), or we must prepare standards ourselves. The second kind of difficulty is more specific of the material analyzed, and is due to the dispersity of molecular weights and structures as well as to the micellar nature of· solutions of bituminous compounds. One of the fundamental hypotheses in the quan-

titative treatment of GPC chromatograms of polymers, consists of assuming that the increment of refractive index is a linear function
of the concentration and is independent of the molecular weight.

However, in the case of asphalts, the increment of the index is
not proportional to the concentration and is not independent of the
elution volume for at least three reasons:

-the presence of compounds of low molecular weight (less than
1000 leads, even within the same homologous series, to variation of the increment of the index with molecular weight,

-the structural heterogeneity leads to an average value of
the increment of the index which is a function of composition,
and this latter varies right through the size distribution
spectrum,

-the relatively intense color of the eluate, may cause absorption of some of the light energy normally required to determine the refractive index. This leads to under-estimating
the latter.

We thus see that any attempt to quantitatively interpret a GPC
chromatogram of a bituminous compound, cannot be made without the
experimental construction of two fundamental calibration curves.
These are:

-the variation of the response coefficient of the detectors
as a function of elution volume, which enables signal heights
to be converted into quantities of material,

-the curve establishing the relation between elution volume
(or distribution coefficient) and molecular weight.

Such curves can be obtained only by analysis of narrow fractions isolated by preparative GPC.

Fractionation of Asphalts by Preparative GPC

The classical conditions for a typical preparative fractionation are given below:

-Apparatus: Waters Chromatoprep 101

-Columns: 2 columns (10^3 and 10^4 Å), diameter 5 cm, length
120 cm

-Solvent: Redistilled chloroform

-Flow rate: 20 ml/min

-Injection: 80 ml of 10% solution

-Volume of fractions: 125 ml

The efficiency of the fractionation is estimated by the injection, in analytical GPC, of 0.3 mg of the most representative fractions. Figure 8 shows the chromatograms of fractions 6 to 18. It
is seen that, apart from the fractions of high molecular weight which
have a special behavior (fractions 6, 7 and 8) and which represent
less than 1% of the original substance, there is a good preparative
separation according to molecular size.

Determination of Correction Curves of the Detectors

Experimental determination of the correction curves may be done
in two ways, e.e. either by measurement of the response coefficient
made my injecting narrow fractions directly into the detector, or by
analysis of these narrow fractions, via the chromatographic system,
and interpretation of the surface under the chromatographic peak
relative to the quantity injected.

Injection of Narrow Fractions Directly into the Detectors

The preliminary tests show that injection, by means of a six-
way valve inserted in the circuit between the columns and the detectors, of 1.5 ml of a solution of known concentration under a solvent
rate of 1 ml/min, gives a signal with a plateau which is sufficiently well-defined that its height is of the same order as the response
of the detector for the concentration of solution examined. Each
fraction from preparative GPC is characterized under the conditions
mentioned above. As far as concerns the refractometric detector it
is seen that all the fractions, except that corresponding to the
substances with the highest molecular weights, give linear relationships if the sensitivity is at least equal to that generally used
(8X). It is not surprising that the range over which there is linearity is smaller in the case of substances with large molecules,
since these are the most highly colored.

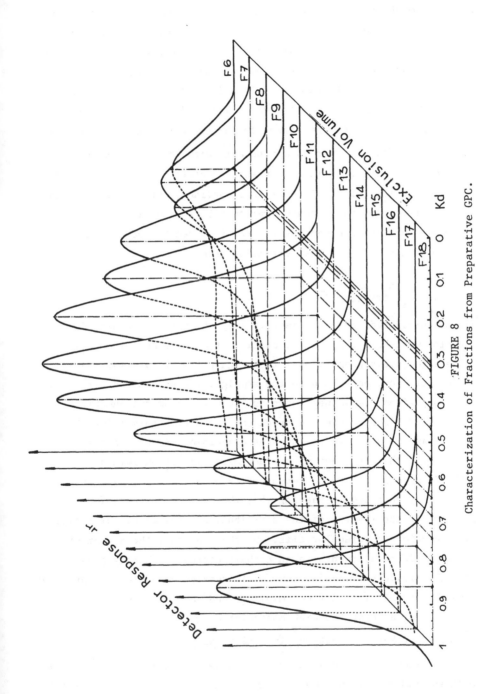

FIGURE 8

Characterization of Fractions from Preparative GPC.

Figure 9 shows the curves obtained for three sensitivities of the refractometric detector for the fraction with the highest molecular weight (fraction 7; sensitivities 16X, 8X and 4X), and the lines corresponding to the usual sensitivity (8X) for the other fractions.

The response coefficinet of the differential refractometer, for a given fraction, is equal to the angular coefficinet of the line relating the height of the signal to the concentration. It is seen that this coefficient varies appreciably with the fraction and therefore with the elution volume. An angular coefficient may be assigned by measuring the height of the signal obtained, for the sensitivity 8X, with the solution of concentration C/32 (C = 0.5%). Moreover, if we assume that the average coefficient is given by the line obtained for the original asphalt, a correction term

$$k_i = \frac{h \text{ asphalt}}{h_i \text{ (fraction i)}}$$

may be calculated for each fraction i by which the response of the detector must be multiplied to obtain, at an elution volume V_i, a value effectively proportional to the quantity of material. A similar calculation, performed on the straight lines obtained for the UV detector, gives the values shown in Table 2. From the numerical values in Table 2 we can construct the correction curves for the detectors as shown in Figure 10. The dispersion of the experimental points from the differential refractometer may be explained by the lack of reproducibility of measurements made with this detector.

It may be noted in passing that the correction curve for the UV detector is the inverse of the distribution curve of polyaromaticity as a function of molecular size. We thus see that the amount of polyaromatic constituents in an asphalt decreases sharply passing from the smallest molecules towards increasing molecular weight, goes through a fairly well-defined minimum, and then increases again to reach a plateau near the high molecular weights.

Injection of Narrow Fractions in Analytical GPC

The correction curve of the detectors may be otained by analysis with analytical GPC of narrow fractions isolated by preparative GPC.

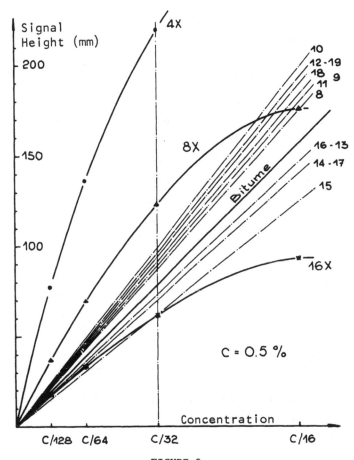

FIGURE 9
Relation between the Height of the Refractometer Signal and Solution
Concentration for Fractions from Preparative GPC.

For this purpose a known volume of a solution of each fraction, of
fixed concentration, is injected. Analysis of the chromatograms
permits determination of the area under the peak from each detector.
The response coefficient is then calculated from the expression:

$$q_i = S_i / m_i$$

where

q_i is the response coefficient of the detector considered for the
i th fraction

TABLE 2

Construction of Detector Correction Curves.

Fraction	Kd	h_i Refract. (mm)	k_i (Refract.)	h_i U.V. (mm)	K_i (U.V.)
7	0.332	123	0.642	108	0.435
8	0.347	88	0.898	128	0.367
9	0.405	93	0.849	120	0.392
10	0.455	100	0.790	111	0.423
11	0.514	91	0.868	102	0.461
12	0.563	97	0.814	71	0.662
13	0.611	74	1.068	53	0.887
14	0.664	70	1.129	35	1.343
15	0.710	63	1.254	27	1.741
16	0.739	74	1.068	24	1.958
17	0.755	70	1.129	28	1.679
18	0.774	97	0.814	33	1.424
19	0.829	97	0.814	54	0.870
Asphalt	-	79	-	47	-

S_i is the area under the peak obtained with the detector considered for the i th fraction

m_i is the quantity injected.

The average response coefficient is then given by the expression
$$Q = \Sigma S_i / \Sigma m_i$$
and the correction term which permits conversion, for each fraction,
of the value of the signal from the detector into quantity is then
given by
$$k_i = \frac{Q}{q_i}$$

Figure 11 makes possible a comparison of the results obtained
by this type of determination, with the curves established by direct
measurement of the response coefficient of the detectors, this being
in the case of the same asphalt Elf Feyzin 80/100. There is fairly

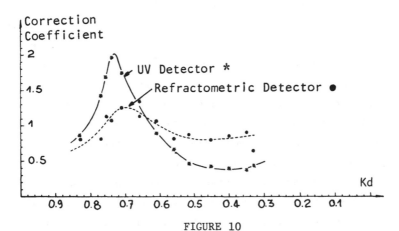

FIGURE 10

Correction Curves of Detectors Obtained by Direct Measurement of
Response Coefficients.

good agreement between the correction curves resulting from the two
techniques in the middle part of the graph. At the ends, the curve
obtained by injection of the fractions into the chromatograph is
clearly above the other curve. This phenomenon may be the preferential adsorption of polar substances, which have been shown to be
present at the ends of the distribution spectrum in a complimentary
study on the characterization, by high performance liquid chromatography, of fractions from preparative GPC[12].

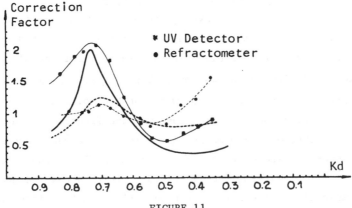

FIGURE 11

Figure 12 shows the experimental points corresponding to the
two 40/50 asphalts (Shell and CFR), and to the two 80/100 asphalts
(Elf Feyzin and Elf Grandpuits). Although there is a relatively
high dispersion, an average curve may be determined for each detector independent of the type of asphalt in the region of low molecular weights. Beyond a certain value of molecular size corresponding
to a distribution coefficient of the order of 0.5, the correction becomes more specific of the penetration index of the sample.

Figure 13 gives an example of the correction, taking account of
the experimental curves established above, of the chromatograms obtained by UV and refractometry for the asphalt Elf Feyzin 80/100.
It is seen that the corrected chromatograms are very close to one another, which confirms that the method of correction is well-founded,
excepting towards the large molecular sizes for which the determination of correction coefficients becomes more hazardous. We shall
therefore assume that the real GPC chromatogram of the asphalt, i.e.
the differential curve giving the relation between the quantity of
material and the elution volume, corresponds to the average of the
two corrected curves as represented by the continuous thick line of
Figure 13.

RELATION BETWEEN ELUTION VOLUME,

MOLECULAR WEIGHT AND HYDRODYNAMIC VOLUME

Molecular Weight by V.P.O. and Membrane Osmometry

Determination of number average molecular weights, either by
V.P.O. or membrane osmometry, of asphalt fractions obtained by preparative GPC, makes it possible to establish an experimental relationship between elution volume and molecular weight. The majority of
the measurements were carried out with the Mechrolab 301A V.P.O.
Some determinations were carried out, for the fractions of high
molecular weight, with a membrane osmometer of the same manufacturer.

Measurements are carried out in benzene at $37^{\circ}C$, the concentration of solutions varying from 2 to 0.5%. In this way the number-
average molecular weights for the preparative GPC fractions from
four asphalts, namely the two 80/100 samples (Elf Feyzin and Elf

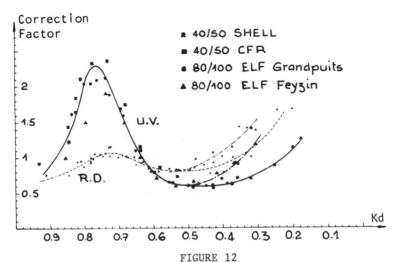

FIGURE 12
Correction Curves of Detectors for Four Road Asphalts.

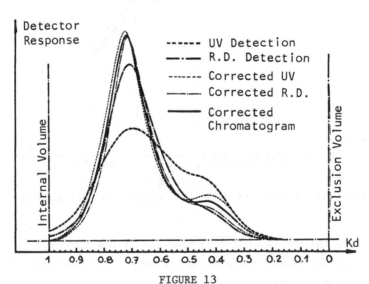

FIGURE 13
Correction Chromatograms Obtained from UV and Refractometry for a
80/100 Asphalt.

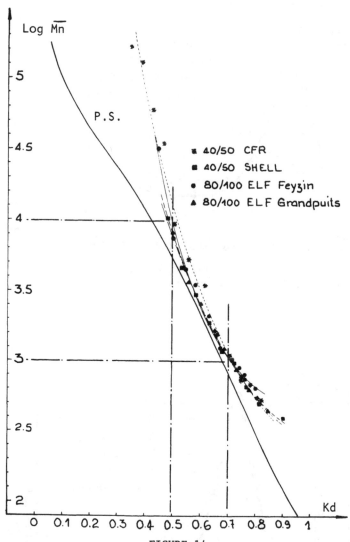

FIGURE 14
Calibration Curve of Molecular Weight for Road Asphalts.

Grandpuits) and the two 40/50 samples (Shell and CFR), were determined. The points representing molecular weight as a function of distribution coefficient are shown in Figure 14. The values for the
molecular weights corresponding to the points on the upper part of
the curve ($\overline{M}_n$ greater than 10,000) were obtained by membrane osmome-

try. On this figure is also shown, as a continuous thick line, the
calibration curve for molecular weight of polystyrenes.

It is seen that, although obtained from four samples of asphalts
for which the origin of the crude and the manufacturing process are
different, the experimental points describe a single curve. This
molecular weight calibration curve, which is specific for asphalts,
is very close to the calibration curve for polystyrenes for distribution coefficients between 0.5 and 0.7. Beyond these values, the
curve for asphalts moves away either to greater elution volumes for
the same molecular weight, or to greater molecular weights for a
given elution volume. It may be noted that the lower part of this
curve (K_d > 0.7 i.e. M < 1000) corresponds to a considerable increase
of aromaticity in substances of low molecular weight (Figure 10);
the increasing deviations from the calibration curve may thus be explained by the superposition of an adsorption process, which is the
stronger the more aromatic are these substances and the lower is
their molecular weight. For values of K_d less than 0.5 (i.e. M >
10,000), we shall see from viscosity measurements that the discrepancy may be explained by the fact that the molecular weight is
measured in the associated state, and not on ideal solutions.

Intrinsic Viscosity of Asphalt Fractions

The sample analyzed is the asphalt CRF 40/50, which we fractionated under the classical conditions. The viscosity measurements were
carried out in THF at 30°C by means of a FICA automatic dilution viscosimeter. Figure 15 shows the experimental points corresponding to
the variation of the expression $(t-t_0)/t_0$ C as a function of the concentration, C, of the solution (t being the time of flow of the solution and t_0 the time of flow of the solvent). We see that, unlike
the behavior observed in the case of ideal solutions, the relationships are not linear for the high molecular weight fractions except
beyond a concentration of the order of 2 to 3 g/l. Below this value,
the curves bend lower down the higher the molecular weight.

This illustrates the dissociability of species as a result of
dilution, and the existence of a critical concentration beyond which
the solutions have a "normal" behavior. The values of intrinsic vis-

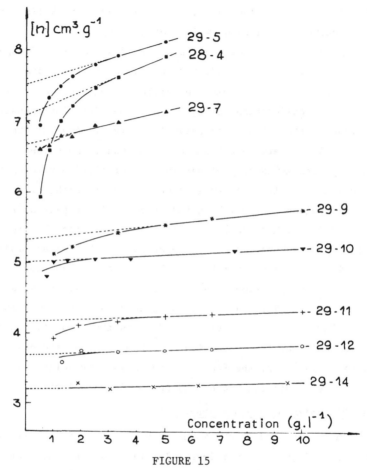

FIGURE 15

Viscosimetric Curves for Asphalt Fractions.

cosity determined by extrapolation, to zero concentration, of the
linear part of the curve are thus apparent values which take account
of an associated state.

Whereas the molecular weights measured by V.P.O. or osmometry
were determined with solutions of concentrations greater than the
critical concentration indicated by viscosimetry, we may conclude
that they also are apparent molecular weights which take account of
the association of species. This experimental fact explains why the
GPC calibration curve of molecular weight for asphalts, moves away
very sharply from the curve for polystyrenes.

Construction of the Curve [η]M for Asphalts

The numerical values of intrinsic viscosities and molecular weights make it possible to construct the curve

log [η]M] f(Kd)

for fractions from the asphalt CFR 40/50. This curve is reproduced in Figure 16, together with the universal calibration curve (continuous thick line). The dotted straight line is the calibration curve of [η]M suggested by Reerink and Lijzenga[11] for asphaltenes. The discrepancy between the curve of [η]M established for asphalt fractions and the universal calibration curve, is qualitatively of the

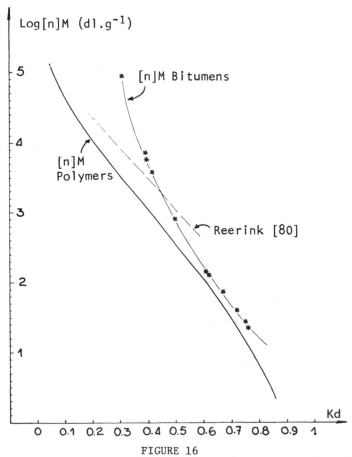

FIGURE 16

Calibration Curves of [η]M for Asphalt Fractions.

same order as that which was observed in the calibration of molecular
weight. It may be considered that the deviation observed in the lower part of the curve is due to the elution lag caused by adsorption
of aromatic substances of low molecular weight. For fractions with
a distribution coefficinet less than 0.5, we have shown that the
numerical values of intrinsic viscosity and molecular weight were
apparent values greater than the real values. This is because
measurements were not carried out on ideal solutions, but on solutions of concentrations greater than the critical value below which
there is dissociation of micellar agglomerates. The product [η]M
value thus measured is therefore greater than that which is obtained
in GPC.

Calibration of Molecular Weight of Asphalts by use of the Universal Calibration Curve

Because the micellar behavior of solutions of asphalt fractions
of high molecular weight leads to apparent values greater than the
real values in determinations of molecular weight by osmometry, it
seems difficult to construct a calibration curve of molecular weight
for asphalts by the traditional methods. Under these conditions we
have assumed, as a working hypothesis, that the universal calibration
of Benoit et. al.[13] applies to asphalt fractions.

The use of universal calibration, in the case of GPC on micro-
packings, requires continuous measurements of the viscosity of the
eluate. It was Ouano[14] who first replaced the measurement of flow
time by that of the pressure drop across a capillary tube. The principle was taken up by Lesec[15], who showed that measurement of the
pressure at the inlet of a capillary tube placed between the outlet
of the columns and the inlet of the detector, could be interpreted
in terms of viscosity.

We therefore assembled the viscosimetric detector described by
Lesec, consisting of a capillary tube of diameter 0.23 mm and length
3 m. A connection at the inlet of this capillary allows measurement
of the pressure. A data collecting system permits interpretation
and treatment of the information from the detectors with an automatic electronic calculator.

After calibrating the chromatographic system in terms of hydrodynamic volume by means of polystyrene, we fractionated the asphalts
(4 samples of 180/220, 2 of 80/100, 2 of 60/70 and 2 of 40/50) under
the usual conditions for preparative GPC. Each fraction was analyzed
by analytical GPC with viscosimetric detection so as to determine its
intrinsic viscosity and its molecular weight by reference to the universal calibration curve. Also, we used V.P.O. to determine the
number-average molecular weights for the fractions with small molecules (M.W. less than 10,000. All the measured points are shown in
Figure 17 on which we have reproduced the calibration curve for polystyrenes (continuous heavy line) and the calibration curve of molecular weight for asphalts established earlier by osmometry (continuous
thin line).

It is seen that these measurements make it possible to define a
new calibration curve of molecular weight for road asphalts, a calibration curve which is independent of the origin of the crude and of
the manufacturing process. This curve is much more realistic than
than established by osmometry, as it is constructed by measurement
of the intrinsic viscosity by GPC with viscosimetric detection.
However, its validity is limited by the hypothesis initially made of
the applicability of the universal calibration to asphalt fractions.

It is also seen that this curve moves away from the curve for
polystyrenes the more the molecular weight increases, and this tends
to show the growing compactness of the heavy fractions from asphalts.

CONCLUSIONS

The experiments described in this report show that GPC on micro-
packing constitutes a physico-chemical analytical tool which can
make a fruitful contribution to the difficult problem of characterization of asphalts.

Qualitatively, it may be noted that the method makes it possible
to obtain, within ten minutes or so and under our working conditions,
a precise and reproducible picture of the apparent distribution of
molecular size in the entities constituting the binder. However,
this reproducibility is not actually attained in practice unless all

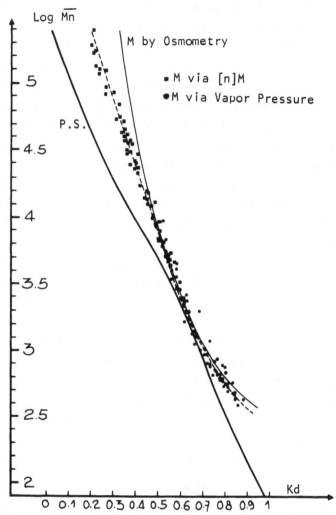

FIGURE 17

Calibration Curve for Asphalts from [η]M.

the operating parameters and, in particular, the concentration of the
solution injected, the volume injected, the nature of the solvent, its
flow rate, the time interval which elapses between preparation of the
solution and its injection, etc. are rigorously defined.

The experimental study of the problems presented by the quantitative interpretation of the chromatograms reveals two kinds of difficulties:

The first is related to the inability of the detectors, whatever
type they may be, to give a response which is independent of the elution volume. This is due to the structural heterogeneity of the con-
stitutents in asphalts, and to the fact that the relationship between
these different chemical constituents is a complex function of molecular size. The quantitative interpretation of a detector signal can
therefore be made only if its correction curve is known. These curves
may be constructed by injection of narrow fractions of asphalts, isolated by preparative GPC, either directly into the detectors or at the
inlet of the analytical system. We have shown that these curves vary
appreciable from one detector to another (differential refractometer
and ultraviolet detector), but not very much with the type of asphalt
studied.

The second kind of problem consists of assignming a molecular
weight to each elution volume. We determined, for various asphalts,
a specific calibration curve relating the number of average molecular
weight of narrow fractions, measured by V.P.O. or osmometry, to their
elution volume (or their distribution coefficient). These curves
are, to all intents and purposes, superimposable, and permit definition of an average curve independent of the origin of the asphalt
and its process of manufacture. It is found that the calibration
curve of molecular weight of asphalts is very close to that of polystyrenes for values between about 1000 and 3000. For lower molecular weight, it moves away towards greater elution volume because of
the increase of aromaticity in the asphalt constituents; increasing
aromaticity with decreasing molecular weight accentuates adsorption
phenomena. When the molecular weight increases, it also moves away

in the same direction; measurements of the molecular weight by V.P.O.
and osometry lead to apparent values with take account of a greater
degree of association than that obtaining in GPC analysis.

In order to offset this difficulty we made the assumption that
the universal calibration is applicable to asphalt fractions, and
used a viscosimetric detector to construct a new calibration curve
of molecular weight by reference to the calibration curve of hydro-

dynamic volume. This last curve, which is specific for asphalts, is also independent of the origin of the crude and of the manufacturing process.

REFERENCES

1. Altgelt, K.H., Die Makrom. Chem., 88, 75 (1965).

2. Altgelt, K.H., J. Appl. Polym. Sci., 9, 3389 (1965).

3. Breen, J.J., Stephens, J.E., Proc. Ass. Asphalt Technol., 38, 706 (1969).

4. Richman, W.B., Proc. Ass. Asphalt Paving Technol., 36, 106 (1967).

5. Bynum, D., Traxler, R.N., Proc. Ass. Asphalt Paving Technol., 39, 683 (1970).

6. Albaugh, E.W., Talarico, P.C., Davis, B.E., Wirkkala, R.A., A.C.S. Symposium on GPC, Houston, 1970.

7. Dickson, F.E., Wirkkala, R.A., Davis, B.E., A.C.S. Symposium on GPC, Houston, 1970.

8. Dickson, F.E., Davis, B.E., Wirkkala, R.A., Anal. Chem., 41, 10, 1335 (1969).

9. Dickie, J.P., Yen, T.F., Anal. Chem., 39, 14, 1847 (1967).

10. Snyder, L.R., Anal. Chem., 41, 10, 1223 (1969).

11. Reerink, J., Lijzenga, J., Anal. Chem., 47, 13, 2160 (1975).

12. Such, C., Brule, B., Baluja-Santon, C., J. of Liq. Chrom., 2(3), 437 (1979).

13. Benoit, H., Grubisic, Z., et al, J. Chim. Phys., 63, 1507 (1966).

14. Ouano, A.C., J. Polym. Sci., A1, 10, 2169 (1972).

15. Lesec, J., Quivoron, C., Analusis, V4, 10, 456 (1976).

AUTHOR INDEX

Author's names are followed by page numbers and, in parentheses, reference numbers.

SUBJECT INDEX

257